Cell Signalling

Cell Signalling

NOEL G. MORGAN

THE GUILFORD PRESS
New York   London

Published in the USA and Canada by The Guilford Press
A Division of Guilford Publications, Inc.
72 Spring Street, New York, NY 10012

This book is printed on acid-free paper.

Last digit is print number: 9   8   7   6   5   4   3   2   1

Library of Congress Cataloging-in-Publication Data

Morgan, Noel G.
  Cell signalling.

  Bibliography: p.
  Includes index.
  1. Cell interaction   I. Title.
QH604.2.M67   1989   574.87   89-11931
ISBN 0-89862-518-1

Printed in Great Britain

for Margaret, Kate and Holly

Contents

Foreword

The fields of cellular communication and signal transduction have moved so rapidly in recent years, that there are few who can consider themselves experts in other than specific areas within these fields. Those working on cellular signalling processes encompass almost all areas of the basic life sciences including biochemistry, cell biology, molecular biology, physiology, pharmacology and biophysics. Although most who study signalling mechanisms are trained within one of these disciplines, it is naïve to consider that a single approach or area of study can be addressed in isolation. Technical advances in methods for measuring the activities of signalling proteins and molecular approaches to determining the structure-function relationships of such proteins have moved exceedingly rapidly within each of these disciplines. Indeed, it is difficult to follow the extensive current literature thoroughly enough to remain completely informed of these advances. Considering these facts, it is crucial that concise information be available on the multitude of signalling mechanisms that function within cells. *Cell signalling* represents an important assimilation by Dr Noel Morgan of the recent advances within this area, and is of great value to workers in each of the varied disciplines that have contributed to our understanding of signalling mechanisms.

Dr Morgan's background and experience has enabled him to write a remarkably comprehensive and up to date text on the fundamental areas of cell signalling and the mechanisms by which receptors are coupled to cellular activation. The book contains a well balanced and clear description of each of the major sub-areas of receptor-induced signalling mechanisms in cells. It is written to provide an original and somewhat different perspective on cell signalling, centering on the second messenger and mechanistic basis of signalling pathways. In other works that have attempted to cover the broad area of cellular receptors, the more usual approach has been to group receptors by the chemical composition of their effector molecules. This has had the effect of turning such books more into endocrinological texts. Dr Morgan has instead centred on the different classes of signalling processes that exist in cells, and has provided very complete descriptions of each of these systems. It is important also that the book contains much conceptual information on the function of cellular signalling mechanisms. Thus, the early sections of the book on principles of cellular signalling, on the theory and regulation of receptors, and on the transduction of information by G-proteins, represent a remarkably informative and original description of the significance of cellular signalling mechanisms. Later chapters deal in depth with highly topical aspects of those receptors that modulate levels of cyclic nucleotides, metabolism of inositol phospholipids, and the activity of tyrosine kinase.

Perhaps the most crucial aspect of cell signalling is that the responses by cells are an integrated assimilation of a large array of signals arising from different external messengers and propagated via distinct second messengers within cells. Thus, cellular signalling pathways are by definition highly interrelated with one another and are rarely effective in isolation without cross-modification. For this reason, it is essential that those researchers predominantly interested in one particular signalling pathway be well aware of the details and nuances of other signalling mechanisms that may be interconnected. As the molecular structures become available for many of the components of the signalling pathways, including receptors, G-proteins, effector enzymes, channels and kinases, remarkable evolutionary relationships between the molecules become apparent. Functional correlates between seemingly unrelated proteins in such pathways become all the more apparent to those investigating structure when such workers are aware of the precise molecular mechanisms involved in parallel systems. Thus, it has become clear that many distinct receptors functioning via different second messenger systems have very similar molecular structures. Moreover, remarkable structural analogies exist between effector enzymes and channels, which although highly distinct in terms of function and mechanism, share in common the ability to be activated via G-proteins. Indeed, the G-protein family is now understood as playing a pivotal role in signal transduction, mediating a multitude of receptor-activated events, and being crucial in determining the effectiveness and sensitivity of signalling processes. Knowledge on their varied roles both in mediating external receptor-induced events within the plasma membrane as well as their control of intracellular signalling and trafficking events on internal membranes deep inside cells, will doubtless continue to accelerate in the future.

The immense interest in these fascinating and fundamental control systems in cells has resulted in the field of cell signalling attracting an enormous number of workers. It is probable that most of these, including researchers at many different levels, will benefit from Dr Morgan's book. Thus the book has a wide application, from students interested in cell signalling mechanisms through to advanced researchers in many of the disciplines applied to such study. The level of interest ranges from those who are undertaking formal study in many areas of biochemistry, physiology, and pharmacology, to those who are working on specific signalling mechanisms and who, by virtue of the rapid development of the field, wish to keep current with parallel systems. It is likely that all who read the book will benefit greatly from its perspective, balance, and quality.

Donald L. Gill
Associate Professor
Department of Biological Chemistry
University of Maryland School of Medicine

1 Hormones and cell signalling

Principles of cellular communication

It is an inescapable fact that all cells must engage in some form of communication even if this is only a rather rudimentary 'one-way' mechanism which provides the cell with the facility to detect and respond to environmental stimuli. When cells become organized together into functional groups (e.g. to form an organism), it is even more evident that individual cells need to be able to sense both the general status of the whole organism in relation to its environment, and the particular functional status of other cells. In fact, it is clear that multicellular organisms could never have evolved without the simultaneous development of appropriate intercellular communication systems.

Some of the mechanisms that cells use to influence one another rely on actual physical contacts, which allow the formation of intimate junctions between neighbouring cells. An example of this is the 'gap junction' which forms a pore between two cells whose plasma membranes are in direct contact, and which facilitates the direct exchange of cytosolic constituents (e.g. ions and low-molecular-weight metabolites) between the cells. Since each cell in any given tissue can form gap junctions with several of its neighbours and these can then also form similar links with other cells, it is possible, by this mechanism, for information to pass quite large distances within a tissue or organ. Indeed, this can be demonstrated quite readily by micro-injection of a dye into a single cell and observation of the movement of the dye amongst neighbouring cells.

However, this type of direct-contact communication suffers from several disadvantages, including the necessarily local nature of the response and the rather slow rate of information flow between the cells. Therefore, organisms have also developed other methods of intercellular signalling which are rapidly propagated and can reach widely distributed tissues. These involve both the nervous and endocrine systems and, although these subserve rather different functions in the organism as a whole, they exhibit remarkably similar characteristics at the molecular level. This may seem surprising, but becomes less so if you consider that the problems imposed upon a cell which is required to detect and interpret an incoming signal are very similar irrespective of whether that signal originated from a cell which is only a few nanometres away (e.g. in a synapse) or from one which is a metre or more away (in the case of endocrine signals). Thus, the molecular details of the signal recognition and transduction processes can be very similar for cells that are participating in

1

entirely different sorts of communication system within a whole organism. For this reason, the distinction between hormonal and nervous inputs is a difficult one to make at a biochemical level, and cells use similar mechanisms to decode both types of signal. Indeed, it is worth stating that many of the molecules that are used to carry information between cells can be found in both the nervous and the endocrine systems in mammals. For example, small peptides such as somatostatin (14 amino acids) can act as both hormones and neurotransmitters, as can the catecholamines adrenaline and noradrenaline.

In this book, the purpose is not to make distinctions between the different systems of communication that exist within multicellular organisms, but rather to emphasize that, from a biochemical viewpoint, these differences are artificial and that different kinds of cells use essentially similar mechanisms to transduce extracellular signals into intracellular responses. Therefore, differentiation between neurotransmitters, hormones, and 'paracrine' (or local) regulators etc. will not be made, but they will all be treated for what they are – extracellular signals which need to be interpreted in the intracellular environment.

Having established this, it must be pointed out that the focus of the following chapters relates to signal molecules that cannot cross the plasma membrane of cells and the discussion concentrates on the particular types of problem posed by this method of information transfer. Biochemists and cell biologists have long puzzled over the details of transmembrane signalling processes, and it will be evident from what follows that there remain more questions than answers in this field of enquiry. However, it is also true that remarkable progress has been made in some areas within recent years, and it is these developments that will be highlighted.

The messengers involved in cellular communication

It is common practice to view endocrine and nervous communication in different terms and to think of the former as a chemical system while conceiving the latter as an 'electrical' system. When considered in terms of the physiology of each system this is certainly true, but such a classification obscures the basic fact that even electrical events are controlled by chemical mechanisms in cells. Thus, the potential difference generated across the plasma membrane of an excitable cell, or that maintained across the inner membrane of mitochondria, represents the physical manifestation of ion movements at the molecular level. Furthermore, the passage of a nerve impulse across a synapse is achieved by the release and subsequent action of a chemical neurotransmitter. Therefore, at the most fundamental level it is evident that all cellular signalling processes are mediated by biochemicals.

Since many hormones and other chemical messengers act on cell-surface receptors (with the notable exceptions of steroid and thyroid hormones), it is necessary for the cell to interpret the presence of the agonist by generating another 'messenger' to propagate the signal through the intracellular environment. Thus, chemical signalling involves the use of a hierarchical system in

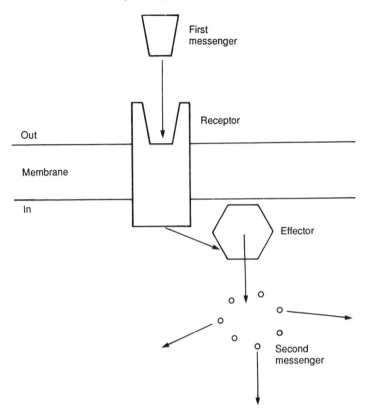

Fig. 1.1 Signal propagation in a hierarchical system

which at least two (and often many more) steps are employed (Fig. 1.1). This introduces a necessary degree of complexity into the signalling process but it also confers several advantages over a more straightforward system.

Signal amplification

The hormone or neurotransmitter that carries information between cells can justifiably be viewed as the first link in the chain of events that control the target cell's response. Hence, this agent is often viewed as the 'first messenger' in the system, and it is responsible for triggering all of the subsequent reactions. By definition, this implies that the next chemical signal generated in response to a hormone represents the 'second messenger' in the process. It is this molecule which sets in motion the intracellular responses to that hormone. It is in the generation of the second messenger that amplification is first introduced into hormone-transduction systems, and this occurs at two levels. In the first place, the initial hormonal input is 'received' by the cell when the hormone

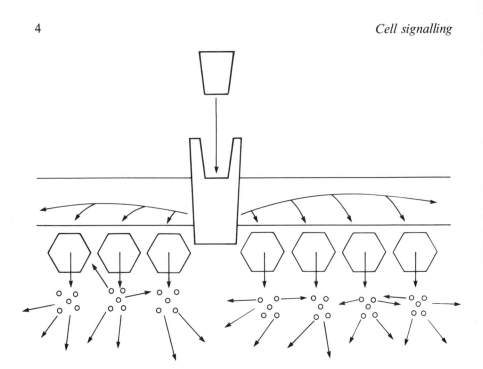

Fig. 1.2   Signal amplification in a hierarchical system

binds to a specific cell-surface receptor. This receptor then mediates the activation of a secondary process to increase the concentration of the second messenger in the cytosol of the cell. In some cases this involves a change in the activity of an enzyme (e.g. adenylate cyclase; phospholipase C), while in others it involves the opening of an ion channel (e.g. nicotinic acetylcholine receptors). In the latter situation there is only one level at which amplification can occur during this step, while in the former case there are two. The reasons for this relate to the mechanisms which many receptors employ to activate intracellular enzymes (considered in detail in later chapters) which involve the use of a separate intermediate group of proteins as signal transducers. Each receptor is activated by a single hormone molecule and can, in turn, catalytically activate several of these transducer proteins. Each of these is then free to promote enzyme activation and second messenger production. To illustrate the immense amplification potential of such a mechanism (Fig. 1.2), it is helpful to consider the situation in which a factor of 10 is introduced at each stage. Thus:

(a)   1 hormone binds to 1 receptor and leads to activation of 10 transducer proteins;
(b)   1 transducer activates 10 enzyme molecules;
(c)   each active enzyme produces 10 molecules of second messenger.

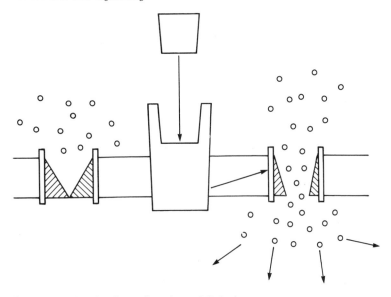

Fig. 1.3 Amplification by an ion-channel linked receptor

Following this calculation through reveals that, in this scenario, the binding of each hormone molecule induces the production of 1000 molecules of second messenger!

In the situation where hormones control the gating of an ion channel, amplification arises in a rather different way but can be equally impressive in magnitude. In this case, the second messenger is not generated within the cell but is primarily located in the extracellular environment. Therefore, a concentration gradient is maintained between the inside and the outside of the cell such that the sudden opening of a channel in the plasma membrane allows a large influx of the appropriate ion (Fig. 1.3). The extent of amplification then depends upon the number of ions that can flow into the cell during the time that each channel remains open. In this context, one ion that is often thought of as second messenger is Ca^{2+}. This is because the concentration of free Ca^{2+} in the cytosol of cells plays a very important role in determining a number of responses, including rates of secretion, glycogen breakdown, motility etc., depending upon the particular cell type. Since much of the Ca^{2+} used to stimulate these various responses derives from the extracellular fluids it is not unreasonable to refer to Ca^{2+} as a 'second messenger'. However, there is a complication in this story since, at present, we do not understand how most of the receptors that promote Ca^{2+} influx into cells achieve this effect. While it is possible that some of them may possess Ca^{2+}-specific channels within their structure (although there is not much evidence for this!) many of them have been shown to promote the activation of phospholipase C with the resultant generation of a water-soluble inositol derivative, inositol 1,4,5-trisphosphate

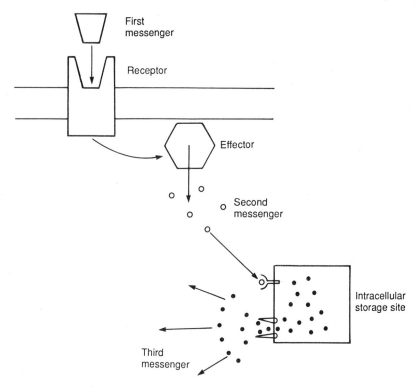

Fig. 1.4 Multiple messengers as features of a hierarchical signalling system

(see Chapter 4). This agent is a true 'second messenger' since it causes the release of Ca^{2+} from intracellular stores. In addition it may also have a role to play in promoting the influx of extracellular Ca^{2+}. Thus, while Ca^{2+} is the active signalling agent which eventually propagates the hormone response in these situations, it is actually the 'third messenger' in the overall sequence of events. Therefore, we may need to extend our classification (Fig. 1.4) beyond first and second messengers to think in terms of still higher orders of messenger molecules! However many messengers we may envisage, the critical point is that their use provides a very efficient means to amplify the hormone signal.

Further amplification also results from events occurring at still later stages of the signal-transduction process which are almost invariably associated with the activation of one or more protein kinase enzymes. Since these have catalytic activity they can alter the phosphorylation state of many substrate proteins to bring about a functional response. Therefore, overall, it can be readily appreciated that a single hormone molecule can cause the generation of many thousands of intracellular effectors in the cytosol of a target cell.

Economy of effector systems

A second advantage to the use of a hierarchy of messengers is that many different hormones all having separate, specific receptors can utilize similar effector systems in their target cells. At first sight, this may appear to be a disadvantage, but, in practice, it provides for great versatility. This is because each individual cell can be equipped with the same type of messenger system (e.g. cAMP or Ca^{2+}) but can respond to the signal which is generated in a unique manner. Therefore, the specificity is introduced at a very late stage of the signal-transduction process and is usually achieved by altering the distribution or availability of the protein kinase substrates from one cell type to another. This allows the various individual cells, which have to respond to hormones in a widely different manner, to utilize identical signalling systems to elicit the appropriate response. If this were not the case, it would be necessary to have an entirely separate system of second messengers for every extracellular signal molecule.

Integration of hormonal inputs

A third valuable aspect of the observed arrangement of hormone–signal transduction systems is that it allows the integration of incoming signals. This is most readily seen when considering the proteins that function as transducers between receptors and effectors and relates to the fact that they display a certain amount of 'cross-talk' between different systems. This idea will be elaborated more fully in Chapter 3 but, simply stated, it means that some of the component subunits may be common to several transducer proteins. The availability of these common subunits in the membrane can then alter the functional capacity of any given protein. For example, one way that some receptors may *inhibit* adenylate cyclase is by altering the state of dissociation of the transducer G-protein involved in *activation* of the enzyme. Hence, in this case, inhibition is not necessarily achieved by any direct effect on the adenylate cyclase catalytic subunit *per se* but results from modulation of the transducer protein that controls this enzyme.

This common example (adenylate cyclase regulation) also highlights another, often overlooked facet of the regulatory process, namely that hormones which act to regulate opposite cellular responses can interact at a very early step in signal transduction if a second messenger is employed. Thus, whereas on the one hand a stimulatory hormone may increase the rate of generation of a second messenger to bring about its effect, an inhibitory hormone can exert its influence simply by reducing the production of that second messenger without the need to directly interact with any other cellular process. For example, insulin can antagonize the ability of glucagon to promote liver glycogen breakdown by inhibiting the production of the crucial second messenger, cAMP. It does not, of necessity have to also inactivate all the other enzymes of the pathway since this will follow automatically from an intervention at the second messenger level.

The system also allows for positive integration between different hormones, although this usually results from effects that occur in the more distal parts of the amplification system. For example, it may occur at the level of the substrates for the various protein kinase enzymes, since in many cases the same target proteins are phosphorylated by both cAMP and Ca^{2+}-dependent kinases. Thus, two independent second messengers can converge at this point.

These considerations emphasize the advantages and versatility of second-messenger cascade systems in cell regulation but also provide pointers to the probable complexity of these processes at the biochemical level. In the following chapters these issues will be considered in greater detail, to provide an overview of our current understanding of the individual components of cellular signalling systems and of the interrelationships which exist between them.

2 Receptors and their regulation

Receptor theory

The idea that hormones or agonist drugs elicit selective cellular responses as a result of interaction with highly specific acceptor molecules has evolved over many years following its initial inception by Paul Ehrlich approximately 100 years ago. Ehrlich was fascinated by the observed selectivity with which some heavy metals are preferentially targeted to the central nervous system and by the reasons that dyes can be used to stain tissues in a selective manner. On the basis of his observations and from a study of immune specificity, he proposed the idea that cells possess on their surfaces defined chemical entities which act as recognition sites for other molecules. This has proved to be a remarkably accurate assertion and it is now almost 'second nature' to modern life scientists to think in terms of agonist–receptor interactions as the starting point of signal generation in cells. Indeed, we now have considerable biochemical information about hormone receptors, including, in some cases, a knowledge of their amino acid sequence, the structure of their functional domains and the mechanisms involved in regulating their functional activities in the cell.

In this chapter, attention will be drawn to certain pharmacological parameters which largely determine the way we think about hormone receptors, and then consideration of some of the recent developments which have furthered our understanding of the structure and function of particular types of receptors.

Hormone–receptor interactions

All hormone molecules whether they be large proteins (e.g. growth hormone) or small lipophilic molecules (e.g. steroids) exert their influence in cells by interacting with target proteins which specifically recognize the hormone. In the case of hormones that can freely pass through the plasma membrane of the cell, the receptor can conveniently be an intracellular protein. This is the case for both steroid and thyroid hormone receptors. However, for those hormones which are either completely unable to, or do not readily cross the plasma membrane, it is clear that the receptors (if they are to be effective) must be located in a position which allows binding of the hormone at an extracellular site. This poses a unique set of problems for the receptor as a functional unit since firstly it must correctly discriminate between the hormone and other extracellular factors and secondly, it must transmit information about the hormone-binding reaction to some other 'effector' system located at a remote site on the opposite side of the plasma membrane. It is this 'transduction' process which is particularly interesting from a biochemical viewpoint and which has proved very difficult to study. However, with the recent advent of gene-cloning techniques coupled with the potential of site-directed muta-genesis, we are now in a position to begin to directly probe the molecular basis of signal transduction by cell-surface receptors.

The pharmacological characteristics of hormone–receptor interactions are likely to be similar for both intra- and extracellular receptors, but the emphasis of this book lies in signal-transduction mechanisms utilized by hormones acting at extracellular sites. Therefore, much of the present consideration of receptors relates to this particular area and should not be taken to imply that similar features are necessarily to be found in intracellular signalling systems.

In recent years, investigation of hormone receptors has largely been carried out with the aid of radiolabelled agonists or antagonists which are used as probes to study the binding process itself. In the case of receptor antagonists, it is obvious that binding parameters cannot be directly correlated with any physiological response (since by definition antagonists do not provoke one) and it is especially important therefore, to ensure that the measured binding site actually represents a physiologically-active receptor site.

The choice of radio-isotope used to probe hormone receptor sites depends upon a number of variables, including the amount of material available and the nature of the ligand molecule itself. If the ligand contains one or more tyrosine residues it can often be radio-iodinated using fairly mild reaction conditions to yield a hormone derivative which can easily be detected due to the emitted γ-radiation. Radio-iodine can be obtained in several isotopic forms, including ^{131}I and ^{125}I, both of which are γ-emitters, and which can each be obtained at sufficiently high specific activity to allow for detection of the small numbers of receptors present on tissue samples. There are, however, some drawbacks to the use of isotopes of iodine, notably their short half-lives ($^{131}I \sim 8$ days; $^{125}I \sim 60$ days) and the possibility that introduction of an iodine moiety into the ligand molecule may somehow alter the recognition process between the receptor binding site and the chosen probe. Nevertheless, radio-iodine remains a widely used 'tag' for the study of hormone–receptor interactions. A second isotope which has also proved useful is ^{3}H, which has a comparatively long half-life (more than 12 years) and can also be obtained at high specific activity (although rather less than the isotopes of iodine). However, it is difficult to label large molecules such as proteins with ^{3}H and consequently its use has largely been confined to small hormones and drugs (e.g. catecholamines and cholinergic ligands).

Saturation binding

Since cells possess a finite number of receptor molecules on their surfaces, it follows that the binding of a particular hormone to these sites will display

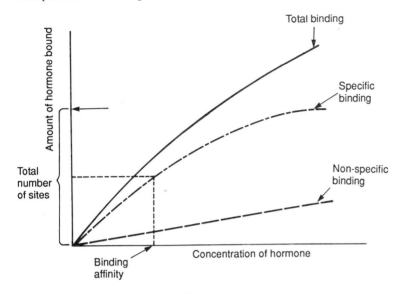

Fig. 2.1   Graph of hormone binding versus concentration

saturation kinetics whereby the extent of binding reaches a plateau as all of the available sites become filled. A plot of the amount of hormone bound versus its concentration will yield a curve from which an affinity term (K_d) can be derived, together with information about the total number of available binding sites (Fig. 2.1). In such experiments account must also be taken of any binding of hormone which occurs at sites that are not true receptor sites but which contribute to the total accumulation of labelled hormone in the tissue. Such binding is usually referred to as 'non-specific' and is estimated by the inclusion of appropriate controls containing a large (~ 100 fold) excess of labelled hormone. This will displace any labelled hormone from true receptor sites but will not alter the non-specific component of the binding. Non-specific binding sites do not exhibit saturation at high concentrations of hormone but binding usually increases in a linear fashion as the hormone concentration is raised.

The binding of a hormone to a receptor site is obviously a chemical process and, as such, is governed by thermodynamic considerations. For example, the binding reaction can be mathematically dissected to yield 'on' and 'off' rates which represent characteristic association and dissociation phases. These are affected by variable factors (such as pH and temperature) which can often alter both phases of the binding reaction. In practice, most binding studies are performed under near-equilibrium conditions which requires that sufficient time be allowed for the achievement of equilibrium. In this respect, it might be imagined that the 'on' rate describing the association of a hormone with its receptor would be extremely rapid, since this is a necessary physiological constraint for hormone action. Indeed, addition of hormones to cells leads to a change in a particular process (e.g. adenylate cyclase activity) within only a few seconds (or less!). However, when binding studies are conducted *in vitro* the measured rate of association is often much slower than would be anticipated from whole cell studies. Indeed, it has been estimated that the binding of some large protein hormones (e.g. growth hormone) to membrane receptor sites can take as long as 24 hours to reach equilibrium. Thus, account must be taken of the detailed time course of hormone binding when measurements of receptor sites are being made.

Scatchard analysis of ligand binding data

A commonly used method for analysis of hormone binding data is one that involves a mathematical transformation of the curve, which relates the amount of hormone bound to its concentration, to yield a linear relationship (Fig. 2.2). The widespread use of this method probably reflects the fact that the transformed data can be manipulated to yield quantitative information about both the affinity of the binding sites for the liquid and the total number of available sites. The method can also provide evidence of any possible heterogeneity within the total population of receptors. The basic mathematics were first worked out about 40 years ago by Scatchard and the manipulation still bears his name, although the commonly used form of analysis is actually a more recent variation on the original method that he proposed. The Scatchard plot is usually displayed in the form:

$$\frac{[\text{Bound hormone}]}{[\text{Free hormone}]} \text{ vs } [\text{Bound hormone}]$$

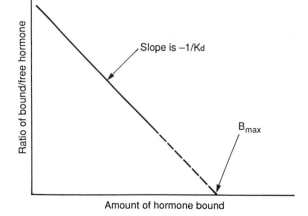

Fig. 2.2 Binding data displayed after Scatchard analysis

In the situation where one type of ligand interacts with a single population of receptor sites, a plot of the data in this form will yield a straight line whose slope is equivalent to $-1/K_d$ and which can be extrapolated to cross the X-axis at a point which is equal to the density of the receptors present (usually referred to as B_{max}). Thus, the data from binding experiments can be used to derive this information very easily.

Positive cooperativity

Unfortunately, Scatchard plots obtained from ligand binding experiments often yield results which cannot be readily joined by a straight line. In general, two basic types of non-linear Scatchard plots may be encountered where the curvature is either convex or concave, respectively (Fig. 2.3). The first of these is probably the less common experimental finding and may be indicative of the existence of a phenomenon referred to as *positive cooperativity*. This means that the binding affinity of the ligand for the receptor population actually increases as more receptor sites become occupied. There are several possible biological explanations for this phenomenon:

(a) Each receptor molecule may possess several hormone binding sites within its structure which can interact in a functional manner. Thus, if one site becomes occupied by a ligand molecule, this will induce a conformational change in the overall molecule such that the binding of ligand to a second site on the same molecule is favoured. Hence, in this situation it is more likely that a new molecule of ligand will bind to a receptor that already contains one ligand molecule than to a completely unoccupied receptor. There is, therefore, a marked positive cooperativity of binding which tends

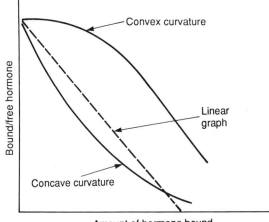

Fig. 2.3 Deviation from linearity after Scatchard analysis

to promote the binding of more ligand molecules to each receptor once the first one has become bound.

This type of interaction can occur in both multi-subunit receptors where the individual binding sites are located on separate subunits, or on single subunit species where the multiple binding sites all occur on the same polypeptide chain.

(b) Individual receptors may be linked together in a functional sense by interactions that take place in the plane of the membrane. Thus, although the receptors exist in the cell as entirely separate molecular species, they must have some form of 'communication' mechanism which allows information about a binding event on one receptor to be transmitted to neighbouring molecules which then undergo a change in conformation, allowing them to bind a molecule of ligand with increased affinity.

A note of caution should be introduced at this point, since there are also several possible artifactual reasons why apparent positive cooperativity might be observed in a receptor population. For example, as discussed above, the introduction of a radiolabel into a ligand molecule can alter its binding properties at a receptor site. If this happens, and a dose–response curve is constructed in which increasing ratios of unlabelled to labelled ligand are used, this could produce a convex Scatchard plot if the affinity of the binding of the labelled species is lower than that of the native ligand.

Negative cooperativity

More commonly, a Scatchard plot of experimental data may yield a curve which is concave rather than convex. This can be interpreted as an indication of the existence of *negative cooperativity* within a receptor population, but more often, it results from the presence in the preparation of more than one type of receptor for the ligand in question, each of which has its own unique binding affinity. In this latter case, it is not necessary to postulate the existence of direct interactions between the receptor molecules since the very existence of multiple binding sites with differing affinities will give rise to a curvilinear Scatchard plot.

True negative cooperativity can arise in a manner similar to the mechanisms described above for positive cooperativity, except that the binding of the first ligand molecule leads to reduced affinity at the functionally-linked binding sites, and produces a situation in which the subsequent binding of a second molecule of ligand is less likely to occur. The insulin receptor is an example of a receptor displaying true negative cooperativity and this is described in more detail in Chapter 6.

A third scenario which can also lead to the production of a Scatchard plot exhibiting the characteristics of negative cooperativity occurs when a given receptor population comprises molecules that exist in several forms having variable affinities for the ligand. This occurs most commonly under conditions where some, but not all, of the available receptors are coupled to another protein which has direct effects on the functional status of the ligand binding site. In particular, the class of proteins called guanine-nucleotide binding (G) proteins (see Chapter 3) intervene to couple many types of cell-surface receptors to intracellular messenger systems. These proteins form a link between the receptor and its effector system and, as such, directly interact with both. In many cases, the interaction with the receptor leads to a decrease in ligand binding affinity. Hence, if some receptors in a population are coupled to G-proteins while others are not (under the conditions in which the binding experiment is performed), this can lead to curvature when the data are expressed on a Scatchard plot.

As in the case of positive cooperativity, any deviation from linearity which is characteristic of negative cooperativity can also be caused by artifacts arising from the design of the experiment. For example, curvature may occur if the ligand molecules form aggregates in solution which leads to inhibition of the binding of the individual component molecules to the receptors; or, alternatively, if the introduction of a radiolabel leads to the formation of a species which has a higher affinity for the receptor than the native ligand.

Hill plot

A second commonly used method of data presentation from binding experiments is based on a mathematical model which pre-dates the Scatchard analysis, and was advanced by Hill as long ago as 1913, to describe the binding of O_2 to haemoglobin. A *Hill plot* expresses (Fig 2.4):

$$\log \frac{[\text{Bound hormone}]}{[\text{Total binding sites} - \text{Bound hormone}]} \text{ vs } \log[\text{Unbound (free) hormone}]$$

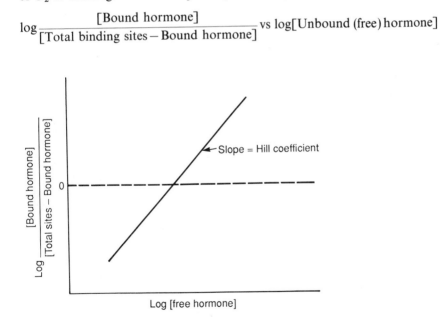

Fig. 2.4 A Hill plot of ligand binding data

and transforms to a linear relationship in which the slope of the line reflects the binding affinity of the receptor for the ligand and is known as the *Hill coefficient*. In a situation where the ligand binds to a single class of exactly equivalent receptors, the Hill coefficient will be 1. A value higher than 1 is indicative of positive cooperativity whereas a value of less than 1 points to the existence of either multiple binding sites or negative cooperativity.

The biochemistry of receptor activation

Our knowledge of receptor pharmacology, as defined in ligand binding experiments, has progressed much farther than our appreciation of how individual hormone receptor sub-types can bring about defined responses in cells. This largely reflects the differences in technology that are required to probe these various events, but it is also true that recent advances in gene cloning have provided new insights into receptor structure and function. Ultimately it should be possible to relate the observed pharmacology of defined classes of receptors to the detailed biochemistry of those receptors. The first steps in this direction have now been taken with the determination of the primary amino acid sequences of some of the receptors which bind catecholamines, acetylcholine, insulin, growth hormone and γ-aminobutyric acid. This has allowed comparisons to be made of the topographical organization of these molecules and has permitted detailed examination of the putative ligand binding and effector coupling regions of the sequences. This analysis has revealed the expected differences in receptor structure and organization, but has also pointed to a surprising degree of homology between receptors which bind different types of ligand, and has suggested that hormone receptors may exist in families whose members share common structural features.

Receptors coupled to guanine-nucleotide binding proteins

The evidence which is now emerging suggests that one particular family of hormone receptors comprises those molecules which interact with their effector systems via G-proteins. These receptors bind a variety of unrelated hormones, ranging from small molecules such as adrenaline and acetylcholine to much larger polypeptides, e.g. glucagon and ACTH. It is surprising, therefore, to find that they may be structurally related molecules. It should be emphasized, however, that at present, we do not have detailed structural data on all of these receptors but the indications from those that have been appropriately analysed supports the view that many of them may be members of a single family.

The prototype for this group of hormone receptors is a molecule that does not, itself, bind hormones but which nevertheless acts as a bona fide receptor. It is the light-sensitive pigment found in the rod cells of the retina, rhodopsin, which consists of the protein opsin bound to a light-sensitive chromophore, retinal. Rhodopsin acts as the 'light-receptor' in the rod cells and its chromophore undergoes a change in conformation in response to the absorption of a photon of light. This reaction is functionally equivalent to the binding of an agonist to a cell-surface receptor. Light absorption facilitates the interaction of rhodopsin with the G-protein transducin (see Chapter 3) and elicits a cascade of events which leads to the eventual propagation of a nerve impulse along the optic nerve. Several genes coding for rhodopsin have been cloned and sequenced and it is thus the best studied member of this receptor family. If a direct comparison is made between the amino acid sequence of rhodopsin and those of the more recently characterized β-adrenergic, α_2-adrenergic and muscarinic cholinergic receptors, it is not immediately apparent that the molecules bear any striking similarities at all. Indeed, the true sequence homology is only about 15% between rhodopsin and these molecules. However, if the putative structural organization of the three types of molecule is compared, the degree of similarity that is revealed is much greater.

Each of the classes of receptor (rhodopsin, β-adrenergic, α_2-adrenergic, muscarinic cholinergic) has within its overall amino acid sequence a series of repeat segments composed of predominantly hydrophobic amino acids. Each of these segments contains between 20 and 28 residues and there are seven such regions within each molecule. It is in these parts of the receptor sequences that the highest degree of conservation is observed; this is evident both between different receptors from the same species and between individual receptors in different species. For example, the degree of amino acid sequence conservation is as high as 95% between some of these regions of the human and hamster β_2-adrenoceptors.

In view of the hydrophobicity of each of the conserved regions it is probable that they represent the membrane-spanning domains of the proteins. Thus, the polypeptide chain must traverse the plasma membrane seven times in each of these receptors. This is probably achieved by the adoption of an α-helical structure within each of the hydrophobic regions (Fig. 2.5), such that the hydrophobic side-chains of the amino acids face the lipid environment of the membrane interior. It is not yet clear why this arrangement has been favoured during the evolution of some types of cell-surface receptor and it is interesting that other receptors exist which do not exhibit this topography. For example, the receptors for insulin and epidermal growth factor (see Chapter 6) have only a single-membrane spanning region in their structures as does the structurally unrelated growth hormone receptor. Thus, the possession of seven transmembrane regions cannot be an absolute prerequisite for effective signal transduction across the membrane. It may, perhaps, be related to the fact that each of these receptors is coupled to a G-protein. Alternatively, it is possible that this arrangement may contribute to the formation of a ligand binding domain in the extracellular space by ensuring the close three-dimensional apposition of regions of the polypeptide chain that are otherwise widely spaced in the sequence. In this context, it can be seen that the presence of seven transmembrane regions results in the formation of three extracellular

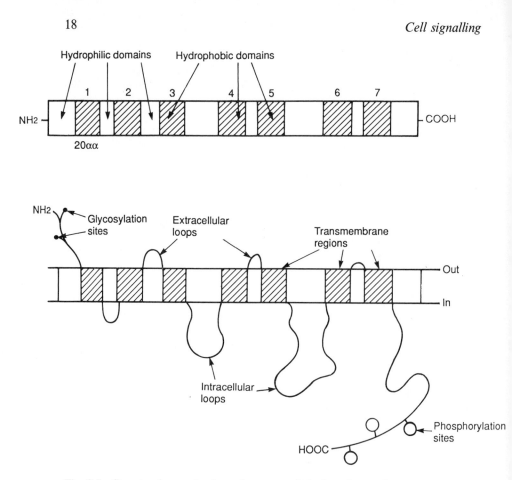

Fig. 2.5 Structural organization of receptors linked to G-proteins

'loops' of polypeptide chain which link each of the hydrophobic segments. These could combine to form the ligand binding domain when folded appropriately. Another possibility is that the membrane spanning regions, together with the extracellular loops, both combine to form the ligand binding site. This idea has gained support from the observation that the hydrophobic regions contain an unusually high number of proline residues which tend to destabilize helical regions by introducing bends into the structure. It is suggested that these bends might create a pocket between the seven helices, into which the ligand is accommodated. Indeed, this arrangement may well represent the situation in rhodopsin where the chromophore (retinal) is held within the hydrophobic core of the protein.

Recently, attempts have been made to resolve some of these possibilities by construction and expression of mutant receptors in cultured cells and examination of their functional properties. These experiments have revealed that, at least in the case of β-adrenoceptors, it is possible to construct mutants that can still bind ligands at relatively high affinity even when large portions of the hydrophilic extracellular domains have been deleted. This suggests that the ligand recognition site may well be formed largely by the association of the hydrophobic membrane-associated regions of the molecule. In support of this, affinity labelling experiments with β-adrenoceptors have revealed that the ligand binding site is closely aligned with certain glycolipid components of the membrane. Furthermore, site-directed mutagenesis has shown that alteration of even a single amino acid residue in the transmembrane region of the molecule can have a profound influence on the extent of ligand binding. For example, replacement of an aspartic acid residue present in the third hydrophobic domain by an asparagine, results in the formation of a receptor which is correctly processed and inserted into the plasma membrane but which is totally unable to bind certain ligands. This suggests that this particular aspartic acid must play a crucial role in the overall binding process, perhaps because it is able to bond with the NH_3^+ group carried by catecholamine hormones. Interestingly, the muscarinic acetylcholine receptor also has an aspartic acid residue in this region of the molecule which may play a similar role (since ligands which bind to muscarinic receptors also have protonated amino agroups) whereas rhodopsin lacks an equivalent aspartate.

Characteristics of the extracellular domain of the receptors

A surprisingly small amount of the total amino acid sequence of G-protein linked receptors appears to be exposed to the extracellular fluids (only 15–20%) but this region contains asparagine residues which carry carbohydrate chains. Thus, the receptors that make up this family are glycoproteins. The total number of glycosylation sites varies for individual molecules (e.g. brain muscarinic receptors have only one site, whereas similar receptors from cardiac muscle have three) and it seems that the presence of carbohydrate is not necessary for either ligand binding or biological activity. The precise role of the sugar residues is not known.

A second major feature of the extracellular domains of all members of this family is the presence of several cysteine residues. This suggests that diverse regions of the structure may well be held together in a defined spatial organization by the formation of disulphide bonds. Indeed the proposal has been advanced that these bonds may play a special role in the activation mechanism (see below).

Characteristics of the cytosolic domain of the receptors

This region of each receptor molecule comprises the carboxy-terminus of the amino acid sequence together with three loops of polypeptide chain that occur between certain of the transmembrane segments. In several members of the group (e.g. β-adrenergic receptors and rhodopsin) there are clusters of serine and threonine residues located near to the C-terminus, whereas these are absent in other members (e.g. α_2-adrenergic and muscarinic receptors). In the molecules where these are present, evidence has accumulated that at least some of them may represent sites of receptor phosphorylation. Several enzymes have been implicated in mediating this response, including a specific β-receptor kinase, cAMP-dependent protein kinase and protein kinase C. Phosphorylation by these enzymes may be part of the regulatory mechanisms involved in controlling agonist sensitivity (see below).

One of the most important functions of the cytosolic region of the receptors is to interact with, and to activate, G-proteins. To date, it has not proved possible to pin-point the precise mechanism by which this occurs nor to identify with certainty which parts of the cytosolic region are involved. However, it has been revealed that the greatest variability between the different members of this class of receptors occurs in the region which forms a cytosolic loop between the fifth and sixth transmembrane segment. It is possible, therefore, that the degree of variability that occurs here may be part of the mechanism by which specificity is maintained. Thus, the structure of this region may possibly determine with which of the available G-proteins any given receptor is able to couple. However, not all of the currently available evidence favours this view, since proteolytic cleavage has been employed to produce a variant of the avian β-adrenoceptor which lacks a large part of the hydrophilic loop between transmembrane regions 5 and 6, and also much of the C-terminal domain. Despite these modifications, the mutant receptor is still able to activate the G-protein G_s at a rate comparable to normal, suggesting that interactions between these components are unaltered by removal of much of the large cytosolic loop. It remains to be established, therefore, precisely how and where such interactions occur.

Mechanisms involved in receptor activation

As indicated above, one striking feature of the structure of G-protein linked receptors is the extreme conservation of cysteine residues. This has been taken as an indication that variations in the oxidation state of critical sulphydryl groups could be involved in the process of receptor activation and signal transduction. In fact it has been shown that treatment of β-adrenoceptors with dithiothreitol (an agent which reduces disulphide bonds) can cause receptor activation under certain circumstances. Examination of the amino acid sequence of β-adrenoceptors reveals that several cysteine residues are present in each of the three distinct receptor domains (extracellular, transmembrane and intracellular), and this pattern is repeated in other representatives of the family. However, the greatest degree of cysteine conservation is observed in those regions of the molecules that reside within the lipid bilayer itself.

If purified β-receptors are electrophoresed under both reducing and non-reducing conditions, it is observed that there is a shift in electrophoretic mobility between the two conditions. The characteristics of this shift are consistent with the possibility that the structure adopted by the receptor protein is more 'compact' under conditions favouring the maintenance of disulphide bridges. This suggests, in turn, that one function of the cysteine residues may be to hold the overall structure together by the formation of intramolecular disulphide bonds, probably between some of the trans-membrane regions.

A provocative observation is that agonists will bind to the compact form of the receptor. It is therefore possible to envisage a model whereby agonist binding leads to a change in either the distribution or the extent of reduction of the intramolecular disulphide bonds which would be equivalent to that induced by thiol reducing agents. Thus, in this view proposed by Malbon and co-workers (1987), the activated state of the receptor would be one in which disulphide rearrangement has occurred such that the various transmembrane regions are allowed to spatially re-organize into a less compact formation. Extensive studies of the effects of thiol reducing agents on receptor activation state suggest that an intermediate form in which only some of the disulphide bonds are reduced may well represent the active form of the receptor, and the proposal has been advanced that agonists induce a similar type of rearrangement in order to promote receptor activation.

It should be emphasized that this model remains speculative at present, but it does provide a working hypothesis which can explain the otherwise enigmatic observation that thiol reagents can cause receptor activation. Of course, it is also possible that this is a coincidental observation and that agonist binding induces a change in conformation which is entirely unrelated to any alteration in disulphide bonding. Whatever the precise mechanism, the extreme conservation of certain cysteine residues between different G-protein linked receptors is strong evidence for an important functional role.

Receptors linked to ion channels

In addition to the recent realization that G-protein linked receptors may all be part of a larger family, it has also become clear that this may be true for a second class of receptors – those which control the permeability of the plasma membrane to ions. This family is entirely separate from that to which G-protein linked receptors belong and includes the nicotinic acetylcholine receptor (in its various forms) as well as receptors for γ-aminobutyric acid (GABA) and glycine, all of which play a role in neuromodulation. The nicotinic acetylcholine receptor controls a sodium ion (Na^+) channel and elicits membrane depolarization when activated, whereas both the GABA and glycine receptors control a chloride ion (Cl^-) channel (probably the same channel in each case) which is inhibitory to neurotransmitter release. Despite this difference in function, it has emerged that these molecules are all structurally related and that they share certain common characteristics.

The basic structure of the nicotinic acetylcholine receptor has been known for several years, as has the fact that this receptor shows species- and tissue-specific variation (at least 20 subunit types have been identified). The receptor which is present in mammalian skeletal muscle is a multimeric protein composed of four types of subunit, each of which is encoded by a separate gene (the structure is $\alpha_2\beta\gamma\delta$). These individual subunits are, however, structurally very similar and each of them is organized into a series of alternating hydrophilic and hydrophobic domains. The organization is such that each subunit has four transmembrane domains within its sequence. The equivalent receptor from the brain has a different subunit composition (only two have been identified) but each subunit also has this same characteristic pattern of hydrophilic/hydrophobic repeat regions.

When the structure of both the GABA and glycine receptors were recently elucidated it became clear that they too exhibit structural similarities to the nicotinic receptor and the existence of this receptor family was first appreciated.

The GABA receptor was purified from bovine brain and found to be a tetrameric glycoprotein composed of two types of subunit, α (M_r 53 K) and β (M_r 57 K). These are arranged as an $\alpha_2\beta_2$ complex in the functional molecule, with the GABA binding site located in the β-subunit. When the α- and β-subunits were sequenced it became clear that the two were very closely related. Both possess an N-terminal signal sequence which is cleaved during protein processing and each subunit also displays an alternating series of domains bearing hydrophilic and hydrophobic amino acids. In terms of strict amino acid homology the two subunits show only 35% identity, although if account is taken of substitutions that do not alter the hydrophilic or hydrophobic nature of the protein, this figure rises to almost 60%. Each subunit also has a large N-terminal domain which is predominantly hydrophilic and possesses several sites at which glycosylation probably occurs *in vivo*. This implies that the N-terminal region is extracellular in the mature molecule.

The most striking feature of the subunit topography is the presence of four regions each containing approximately 22–25 hydrophobic amino acids which probably comprise the transmembrane segments. It is in these regions that the homology with the organization of the nicotinic receptor is most obvious. If the amino acid sequences of the α- and β-subunits of the GABA receptor are aligned with those of the α-subunit of the nicotinic receptor and one of the glycine receptor subunits (only one has been sequenced so far) it is possible to construct a plot which shows the variation in hydrophobic character along each of the sequences (Fig. 2.6). This reveals an extraordinary degree of similarity and provides very strong evidence that each of these polypeptide chains probably adopts a very similar conformation within the membrane environment. This concept is strengthened still further if the amino acid sequences are compared directly, since in some parts of the molecule the sequence homology is as high as 60% (allowing for conservative amino acid substitutions) between the different receptors.

One particularly significant finding is that all of the molecules have a pair of cysteine residues in the N-terminal extracellular domain which may well bond together to form a disulphide bridge and hence a loop in the chain. This region of the polypeptide chain also contains several other highly conserved amino acids and a functional glycosylation site, all of which suggests that the three-dimensional conformation adopted by this region is likely to be crucial to

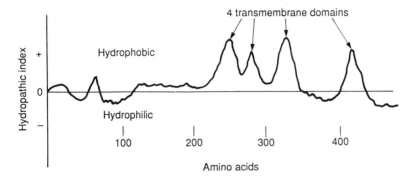

Fig. 2.6 Hydrophobicity plot of a typical subunit from an ion-channel linked receptor

receptor function. A high degree of homology also exists in the transmembrane regions of the different receptor molecules, which may seem surprising bearing in mind that each receptor probably contains a specific ion channel within its structure, and that the receptors do not have identical ion specificities. This suggests that the common features are more likely to be involved in the gating mechanism than in specificity determination. In this context each subunit contains a proline residue in the first transmembrane domain which, as in the case of G-protein linked receptors, is likely to introduce a bend into the membrane spanning helix. It has been suggested that the presence of this feature may result in the displacement of the helix, such that it protrudes into the central 'pore' of the channel and effectively blocks it. This could then form part of the gating mechanism if ligand binding is accompanied by an appropriate movement of the helix, thus allowing the passage of ions.

The second transmembrane domain of all of these receptor subunits also bears homology in that it contains an unusually high number of serine and threonine residues. This is surprising for a sequence which is buried within the membrane, but these amino acids presumably interact to form a suitably hydrophilic environment at the centre of the channel, such that ions can readily penetrate through the lipid environment of the membrane.

On the basis of these considerations a model has been proposed by Barnard and co-workers to explain the functioning of the 'ion-channel receptor' family (Fig. 2.7). It envisages that the transmembrane domains of the constituent subunits are arranged in a pattern which allows them to form the walls of the ion channel in the membrane. The hydroxyl groups present in the serine and threonine side chains face into the lumen of the channel and, when the receptor does not contain bound ligand, the channel is blocked by the 'kinked' helices of the first transmembrane segment of each subunit. This arrangement would facilitate the clustering of a number of charged amino acids at both the intra- and the extracellular faces of the channel which could then form a 'filter' to maintain the specificity of each particular channel (the net charge in these regions will therefore be different for Na^+ and Cl^- channels). Ligands bind to specific sites located in the extracellular N-terminal domains which then alter

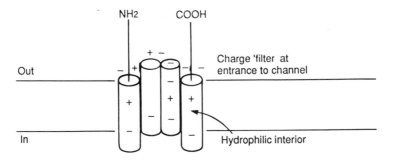

Fig. 2.7   Possible arrangement of each subunit of ion-channel linked receptors

(in a still undefined manner) the charge distribution at the entrance to the channel and the conformation of the bent helix to permit ion flow.

The validity of this model must now be verified in further studies, but it serves to illustrate how diverse receptors controlling the membrane permeability to ions of opposite charge can still be structurally organized in a very similar manner. It also shows that these receptors may all be members of a single family which could have arisen by the gradual evolutionary modification of only a small number of ancestral genes.

Receptor families

The discussion in the previous sections illustrates that as we are learning more about the structural organization of different hormone and neurotransmitter receptors, it is becoming obvious that many such receptors fall into definable categories and that the individual members of any given 'family' simply represent variations on a general theme which is constant throughout that family. This is true for receptors linked to G-proteins or ion channels and to the class of tyrosine kinase receptors, which includes the receptors for insulin and epidermal growth factor (see Chapter 6). However, there are still some receptors that stand alone and seem not to have closely related counterparts. A notable example of this is the mammalian growth hormone receptor, the amino acid sequence of which has now been determined. This protein consists of a single large subunit ($M_r \sim 130\,K$) which bears no obvious structural similarity to any other known protein. Furthermore, the mechanism by which it transmits a signal to the intracellular environment is also a mystery. Perhaps this molecule represents the first example of a new family of hormone receptors, other members of which still await structural elucidation. Whether this is true or not, it serves to remind us that we have only begun to 'scratch the surface' of receptor biochemistry and that there is much work still to be done in this now rapidly developing field of research. No doubt a few surprises are still in store!

Receptor desensitization

It has been realized for many years that the hormone responsiveness of cells is variable and that this represents an adaptive mechanism by which a cell can regulate its state of activation according to the prevailing environmental conditions. This process of adaptation can take the form of both sensitization (in which the cell becomes more responsive to a given concentration of hormone) or desensitization (where hormone responsiveness decreases) and the biochemical mechanisms that underlie these phenomena are now emerging. In the case of cell sensitization, the response is almost always associated with a true increase in the number of receptor molecules available for hormone binding and therefore reflects a net enhancement of receptor recruitment rate. This usually occurs in response to prolonged exposure of a cell to a medium which is deficient in the particular hormone in question, and effectively represents an attempt by the cell to give itself the maximal opportunity to detect any hormone which may be present. In principle, receptor 'up-regulation' can occur by one of two mechanisms:

1. Increased synthesis of new (or redistribution of nascent) receptors.
2. Decreased rates of receptor internalization and degradation.

In practice, both of these mechanisms probably contribute to the overall response. From a biochemical standpoint, the mechanisms involved in receptor desensitization are more interesting and can, potentially, provide greater insights into the control of receptor function. Therefore, focus will be primarily on this aspect of regulation.

Homologous desensitization

Receptor desensitization can occur in one of two forms which differ according to the precise characteristics of the response. In some cases, treatment of a cell with an agonist induces a state of refractoriness in which that cell becomes poorly responsive to a second treatment with the same agonist but remains normally responsive to hormones acting at different receptors. This phenomenon is referred to as *homologous desensitization* since it involves only a single type of receptor. A good example of a receptor which exhibits this phenomenon is the β-adrenoceptor which displays rapid desensitization in response to agonist treatment (Fig. 2.8). Since the β-receptor is coupled to adenylate cyclase via the G-protein, G_s, and all other hormone receptors which stimulate adenylate cyclase also do so via G_s (see Chapters 3 and 4), homologous desensitization cannot be the result of changes in either the cyclase itself or in G_s. If it were, then the response to activation of other receptors would clearly also have to be affected. These considerations suggest that the primary change which takes place must occur at the level of the receptor itself, in order to mediate homologous desensitization. The molecular details of the process are not yet fully understood, but a number of clues have been provided from recent investigations. In the first place, it has been

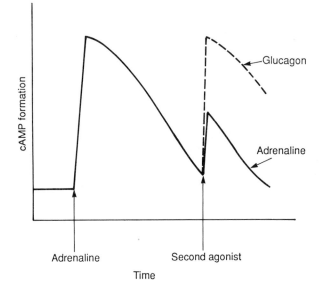

Fig. 2.8 Homologous desensitization of the β-adrenergic receptor in cells responsive to adrenaline and glucagon

established that the desensitization process is initiated very rapidly following agonist binding (within a few minutes) and that it is not accompanied by any physical loss of receptors from the cell surface. Furthermore, the process can be observed in a cell-free system containing membranes, which supports the idea that receptor internalization is not a necessary component. Studies with isolated membranes have also revealed a number of other requirements for the manifestation of homologous desensitization, including the presence of Mg^{2+} and ATP (non-hydrolysable analogues cannot be substituted) and the need for effective coupling of the receptor to G_s. Hence, in mutant cells that lack a functional G_s protein, β-agonists do not induce rapid desensitization.

The requirement for Mg^{2+} and ATP is highly suggestive of the participation of a phosphorylation event in the process and this possibility has been strengthened by the finding that immunoprecipitation of β-receptors from agonist-treated cells yields a form of the receptor that contains labile phosphate. *In vitro*, the β-receptor can serve as a substrate for several protein kinases, including both cAMP-dependent and Ca^{2+}-dependent enzymes as well as protein kinase C. However, there is little evidence to support a role for any of these enzymes in mediation of homologous desensitization in intact cells. The fact that the response is receptor-specific suggests that it is likely to be mediated by a receptor-specific kinase, and just such an enzyme has now been identified in association with the β-receptor. This kinase phosphorylates the receptor only when it contains bound hormone, as would be expected if it plays a role in homologous desensitization. Interestingly, the enzyme can also

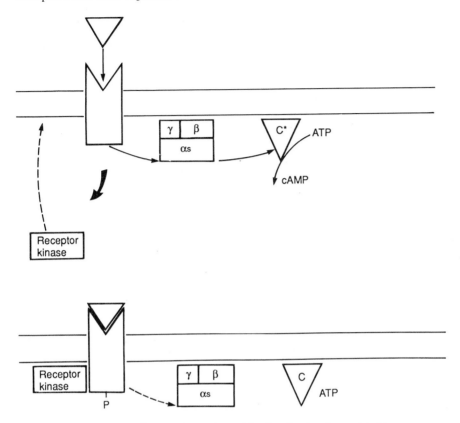

Fig. 2.9 Mechanism of homologous desensitization by the β-receptor kinase. Phosphorylation of the agonist-occupied receptor leads to impaired coupling to G_s and to reduced adenylate cyclase activity

phosphorylate light-activated rhodopsin, an observation which supports the proposal that rhodopsin must share significant structural homology with the β-receptor (see above).

The β-receptor kinase has been purified in homogeneous form and consists of a single protein subunit ($M_r \sim 80$ K) which has a K_m of $\sim 0.25\,\mu$M for the β-receptor. *In vivo* the enzyme incorporates 1–2 molecules of phosphate into each molecule of receptor although much higher levels of phosphorylation can be achieved *in vitro*. The kinase seems to be predominantly located in the cytosol of resting cells but evidence suggests that it is rapidly translocated to its site of action at the plasma membrane when a hormone binds to the appropriate receptor (Fig. 2.9).

The biochemical effect of receptor phosphorylation has not yet been established but, presumably, it must result in uncoupling of the agonist-occupied receptor from G_s. Indeed, this may represent a general mechanism for receptor–effector uncoupling since a unique rhodopsin kinase and an α_1-

adrenoceptor kinase have also been identified, both of which are implicated in desensitization of their respective substrate receptors.

A surprising aspect of receptor phosphorylation has arisen from studies investigating the effect of somatostatin in cultured lymphoma cells. In these, as in other cells, somatostatin binds to a receptor that is negatively coupled to adenylate cyclase (i.e. inhibits the enzyme) and this binding promotes homologous desensitization of the receptor. The surprising result is that this response can be correlated with the translocation and activation of the β-adrenoceptor kinase. Furthermore, kinase activation occurs at the same time as receptor desensitization begins, which is highly suggestive of a link between the two responses. These data suggest, then, that the *same* protein kinase may be able to uncouple both stimulatory and inhibitory receptors from adenylate cyclase.

At first sight this idea does not seem compatible with a mechanism in which agonists provoke the selective desensitization of only that receptor to which they bind. However, since the kinase will only use the agonist-activated receptor as substrate, the specificity is introduced at this point. Thus, in a cell treated with somatostatin it is only the somatostatin receptor that becomes phosphorylated (and desensitized) despite the presence of β-receptors. Conversely, when incubated with β-agonist it is only the β-receptors that become phosphorylated.

On this basis, it can be seen that the same protein kinase can, in principle, control the sensitivity of several different receptors and still produce a state of desensitization which is truly 'homologous' in nature.

Receptor sequestration

Although, in the short term, homologous desensitization is not correlated with any depletion of cell-surface receptors, in some cells agonist treatment does lead to the specific removal of receptor molecules from the cell surface after approximately 30 minutes. In the case of β-agonists this loss of receptors has been correlated with the appearance of ligand binding activity in a vesicular compartment located within the cell. The receptors in this compartment are still accessible to membrane-permeant catecholamine ligands but are not available to solely extracellular hydrophilic ligands. The vesicles in which the receptors become sequestered have been characterized as light-density structures which do not contain either adenylate cyclase or G_s. This suggests that specific compartmentation of the individual components of the effector system represents one part of the process of homologous desensitization. The sequestration mechanism does not lead to immediate degradation of the internalized receptors since full ligand binding capacity can still be recovered from the vesicles even after desensitization has taken place. Furthermore, the sequestered receptors also seem to retain functional activity since they can be recovered and reconstituted into cells such that agonist-induced adenylate cyclase activation is restored. This may seem inconsistent with the idea that the initial event in desensitization results from a phosphorylation-induced uncoupling of the receptor–effector system, which is only later followed by receptor sequestration. However, the data imply that the initial uncoupling reaction must be a transient event and that, by the time sequestration occurs, the process has been reversed (perhaps by a phosphatase). The precise location of the receptors after their removal from the cell surface has not been established, although two proposals have been advanced to explain their redistribution. One suggests that the receptors become incorporated into clathrin-coated pits which then invaginate to form true vesicles, while the second envisages that the vesicles do not actually leave the intracellular face of the membrane, but remain associated with it. Further work is necessary to distinguish between these possibilities.

Down-regulation of receptors

In some cells a third phase of receptor desensitization can also be distinguished in which there is actual loss of identifiable receptors from all compartments of the cell. This usually takes several hours to develop and cannot be readily explained simply on the basis of intracellular sequestration. There is some debate as to whether the process is achieved by complete proteolytic degradation of the receptor population since it has been claimed that in certain cell types, ligand binding activity can be regained without the need for new protein synthesis. Whatever the mechanism, it appears to require functional coupling of the receptors to a G-protein, although the significance of this observation is also obscure. It does not reflect the need to directly activate an effector system since cells which lack, for example, the catalytic subunit of adenylate cyclase still show down-regulation of those receptors that would normally be coupled to this enzyme, even though the binding of hormone to these receptors is unable to elicit enzyme activation.

Receptor endocytosis

Part of the mechanism which leads to homologous down-regulation of receptors involves the complete removal of ligand-bound receptors from the cell surface by an endocytic process. There are several variations on this theme and it can result in receptor internalization, ligand degradation and then recycling of the receptor to the surface, or alternatively both the receptor and the ligand can be degraded. It is only in this latter situation that there is potential for receptor down-regulation since in the case where receptors recycle to the surface, any internalization is rapidly compensated by the re-appearance of the receptors in the plasma membrane. Examples of receptors that are degraded following internalization include those for epidermal growth factor and insulin, and in the case of insulin, this can have important physiological and pathophysiological implications. In situations where circulating insulin levels are maintained above normal (e.g. in obese subjects) the cells of peripheral tissues are exposed to unusually high levels of the hormone and, thus, the proportion of cell-surface receptors which contain bound insulin will be increased under 'resting' conditions. This has the effect of promoting receptor–ligand internalization and leads to the degradation of both species. In the course of time the increased rate of degradation exceeds the rate at which receptors are replaced by *de novo* synthesis and the total receptor population is correspondingly reduced. This, in turn, renders the cell refractory to the hormone and results in a state of insulin resistance whereby the prevailing insulin concentration is not sufficient to efficiently reduce hyperglycaemia. This, therefore, results in still more insulin secretion to boost the already elevated circulating concentration which only serves to exacerbate the situation by facilitating the further down-regulation of the receptors. Hence, this is a self-perpetuating problem which can only be relieved if measures are taken to reduce the persistent hyperglycaemia.

This type of down-regulation by receptor degradation is usually homologous in nature since it only involves the receptor for one particular ligand and, even though other receptors may be present on the cell surface, the rate of internalization of these is unaffected. This mechanism differs from that described in the previous section since it involves the total loss of cellular receptors rather than their redistribution into a compartment which is inaccessible to extracellular ligands.

Mechanism of receptor–mediated endocytosis

Those hormone receptors which can be rapidly internalized upon ligand binding (and this does not happen to all hormone receptors) probably enter the cell by a route which is also employed to internalize a wide range of other extracellular ligands including low-density lipoproteins, immunoglobulins and transferrin. This involves an initial clustering of the receptors into specialized regions of the plasma membrane termed 'coated pits'. These are regions of the membrane which invaginate and become covered on their intracellular face with a semi-crystalline array of molecules composed primarily of the protein clathrin. This is a large protein ($M_r \sim 180\,K$) which spontaneously polymerizes to form a trimeric 'triskelion' structure and produces a characteristic 'fuzzy' appearance when viewed under the electron microscope. The receptors containing bound ligand migrate into these coated pits which then pinch off from the plasma membrane to form vesicles (Fig. 2.10). The receptors are then rapidly (within 2–3 min) delivered from these structures to a second set of vesicles which form part of the endosome, and undergo acidification due to the activity of a proton pump located in the membrane of the endosomal vesicle. It is at this stage that any receptors that are to be recycled back to the cell surface are segregated, and the remainder pass into the lysosomal system where degradation occurs. The molecular basis of the segregation mechanism has not been established but it occurs prior to the fusion of the endosomal vesicle with primary lysosomes, where proteolysis of both the receptor and the ligand takes place.

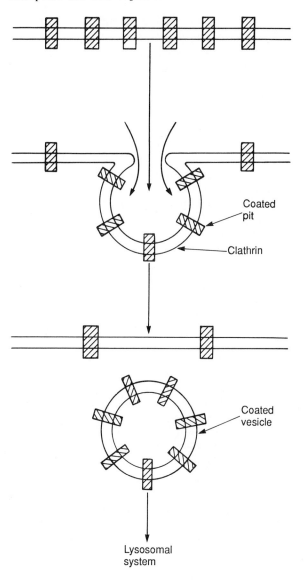

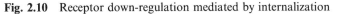

Fig. 2.10 Receptor down-regulation mediated by internalization

Heterologous desensitization

This form of desensitization is characterized by a much more general decrease in hormone responsiveness than is seen in homologous desensitization and involves receptors for more than one type of agonist. In general, it is a slowly developing response which is only manifest after the continuous exposure of a cell to an agonist for several hours. A clue as to the mechanisms involved in this process was provided by the demonstration that agents which artificially elevate cell cAMP levels (e.g. forskolin, cholera toxin) can provoke the heterologous down-regulation of several hormone receptors which are linked to adenylate cyclase. This suggests that the process is a feedback response elicited by the same second messenger system as that which is activated by some of the agonists to which a cell normally responds. One obvious mechanism by which the responses to a range of different hormones all acting on the same effector system could be desensitized involves functional modification of a common coupling protein. In the case of receptors linked to adenylate cyclase this would be G_s. In support of this, functional modification of G_s has been directly demonstrated in cells displaying heterologous desensitization. This was achieved by isolation of the protein and reconstitution into membranes containing adenylate cyclase, but no endogenous G_s. In these situations, the ability of G_s to promote enzyme activation is impaired.

It is not yet certain how G_s becomes modified under these circumstances, although it has been identified as a possible substrate for cellular protein kinases and a change in phosphorylation state is therefore a strong contender.

A change in protein phosphorylation state has also been implicated at another level in the process of heterologous desensitization. Thus, it has been observed in both avian and mammalian tissues that receptor desensitization correlates with the incorporation of ^{32}P into receptors that are linked to adenylate cyclase (e.g. β-adrenoceptors). This reaction is not the same as that described above for homologous desensitization (which is due to the activation of a receptor-specific kinase), since several different types of receptor can become phosphorylated, and the response can be mimicked in cell-free systems by the addition of an exogenous protein kinase, such as cAMP-dependent protein kinase. Interestingly, under these conditions the receptor acts as a better substrate for the kinase if an agonist is also present. Therefore, heterologous desensitization does not require the large-scale removal of receptors from the cell surface and is probably achieved by a combination of two mechanisms. Both receptor phosphorylation and modification of G-protein–effector coupling are likely to be components of the process.

Further reading

Receptor characterization by ligand binding

DeMeyts, P., Roth, J., Neville, D. *et al.* (1973). Insulin interactions with its receptor: experimental evidence for negative cooperativity. *Biochem. Biophys. Res. Commun.*, **55**, 154–61.

Hill, A.V. (1913). The combination of haemoglobin with oxygen and carbon monoxide. *Biochem, J.*, **7**, 471–80.

Klotz, I.M. (1982). Numbers of receptor sites from Scatchard graphs: facts and fantasies. *Science*, **217**, 1247–9.

Klotz, I.M. (1983). Ligand–receptor interactions. What we can and cannot learn from binding measurements. *Trends Pharmacol. Sci.*, **4**, 253–5.

Limbird, L.E. (1986). *Cell Surface Receptors: A Short Course on Theory and Methods.* 196pp. Boston, Martinus Nijhoff Publishing.

Rosenthal, H.E. (1967). A graphic method for the determination and presentation of binding parameters in complex systems. *Anal. Biochem.*, **20**, 525–32.

Scatchard, G. (1949). The attractions of protein for small molecules and ions. *Ann. N.Y. Acad. Sci.*, **51**, 660–72.

Taylor, S.I. (1975). Binding of hormones to receptors. An alternative explanation of non-linear Scatchard plots. *Biochemistry*, **14**, 2357–61.

Receptors linked to guanine-nucleotide binding proteins

Burgen, A.S.V. (1988). The generation of receptor specifity. *Trends Pharmacol. Sci.* **9** (Suppl. 1), 1–5.

Dixon, R.A.F., Sigal, I.S., Rands, E. *et al.* (1987). Ligand binding to the β-adrenergic receptor involves its rhodopsin-like core. *Nature*, **326**, 73–7.

Dohlman, H.G., Bouvier, M., Benovic, J.L. *et al.* (1987). The multiple membrane spanning topography of the β_2-adrenergic receptor. *J. Biol. Chem.*, **262**, 14282–8.

Dohlman, H.G., Caron, M.G. and Lefkowitz, R.J. (1987). A family of receptors coupled to guanine-nucleotide regulatory proteins. *Biochemistry*, **26**, 2657–63.

Gocayne, J., Robinson, D.A., Fitzgerald, M.G. *et al.* (1987). Primary structures of rat cardiac β-adrenergic and muscarinic cholinergic receptors obtained by automated DNA sequence analysis: further evidence for a multigene family. *Proc. Natl. Acad. Sci. USA*, **84**, 8296–300.

Kobilka, B.K., Matsui, H., Kobilka, T.S. *et al.* (1987). Cloning, sequencing and expression of the gene coding for the human platelet α_2-adrenergic receptor. *Science*, **238**, 650–6.

Kubo, T., Fukada, K., Mikami, A. *et al.* (1986). Cloning, sequencing and expression of complementary DNA encoding the muscarinic acetylcholine receptor. *Nature*, **323**, 411–16.

Lefkowitz, R.J. (1988). Adrenergic receptors – models for study of receptors coupled to G-proteins. *J. Biol. Chem.*, **263**, 4993–6.

Malbon, C.C., George, S.T. and Moxham, C.P. (1987). Intramolecular disulfide bridges: avenues to receptor activation? *Trends Biochem. Sci.*, **12**, 172–5.

Marx, J.L. (1987). Receptor gene family is growing. *Science*, **238**, 615–6.

O'Dowd, B.F., Lefkowitz, R.J. and Caron, M.G. (1989). Structure of the adrenergic and related receptors. *Ann. Rev. Neurosci.*, **12**, 67–83.

Rubenstein, R.C., Wong, S.K.-F. and Ross, E.M. (1987). The hydrophobic tryptic core of the β-adrenergic receptor retains G_s regulatory activity in response to agonists and thiols. *J. Biol. Chem.*, **262**, 16655–62.

Strader, C.D., Sigal, I.S., Register, R.B. *et al.* (1987). Identification of residues required for ligand binding to the β-adrenergic receptor. *Proc. Natl. Acad. Sci. USA*, **84**, 4384–8.

Receptors linked to ion channels

Grenningloch, G., Rientiz, A., Schmitt, B. *et al.* (1987). The strychnine-binding subunit of the glycine receptor shows homology with nicotinic acetylcholine receptors. *Nature*, **328**, 215–20.

Karlin, A. (1987). Going round in receptor circles. *Nature*, **329**, 286–7.

Noda, M., Takashi, H., Tanabe, T. *et al.* (1983). Structural homology of *Torpedo californica* acetylcholine receptor subunits. *Nature*, **302**, 528–32.

Schofield, P.R., Darlison, M.G., Fujita, N. *et al.* (1987). Sequence and functional expression of the GABA$_A$ receptor shows a ligand-gated receptor super-family. *Nature*, **328**, 221–7.

Schwartz, R.D. (1988). GABA$_a$-receptor gated ion channels – biochemical and pharmacological studies of their structure and function. *Biochem. Pharmacol.*, **37**, 3369–76.

Stephenson, F.A. (1988). Understanding the GABA$_A$ receptor: a chemically gated ion channel. *Biochem, J.*, **249**, 21–32.

Stevens, C.F. (1987). Channel families in the brain. *Nature*, **328**, 198–9.

Strange, P.G. (1988). The structure and mechanism of neurotransmitter receptors. *Biochem. J.*, **249**, 309–18.

Udgaonkar, J.B. and Hess, G.P. (1987). Isosteric regulation of the acetylcholine receptor. *Trends Pharmacol. Sci.*, **8**, 190–2.

Receptor regulation

Benovic, J.L., Mayor, F. and Somers, R.L. (1986). Light dependent phosphorylation of rhodopsin by β-adrenergic receptor kinase. *Nature*, **321**, 869–72.

Benovic, J.L., Mayor, F., Staniszewski, C. *et al.* (1987). Purification and characterization of the β-adrenergic receptor kinase. *J. Biol. Chem.*, **262**, 9026–32.

Levitzki, A. (1986). β-adrenergic receptors and their mode of coupling to adenylate cyclase. *Physiol. Rev.*, **66**, 819–54.

Mayor, F., Benovic, J.L., Caron, M.G. and Lefkowitz, R.J. (1987). Somatostatin induces translocation of the β-adrenergic receptor kinase and desensitizes somatostatin receptors in S49 lymphoma cells. *J. Biol. Chem.*, **262**, 6468–71.

Sibley, D.R. and Lefkowitz, R.J. (1985). Molecular mechanisms of receptor desensitization using the β-adrenergic receptor-coupled adenylate cyclase system as a model. *Nature*, **317**, 124–9.

Wileman, T., Harding, C. and Stahl, O. (1985). Receptor-mediated endocytosis. *Biochem. J.*, **232**, 1–14.

Other receptors

Leung, D.W., Spencer, S.A., Cachianes, G. *et al.* (1987). Growth hormone receptor and serum binding protein: purification, cloning and expression. *Nature*, **330**, 537–43.

Morgan, D.O., Edman, J.C., Standring, D. *et al.* (1987). Insulin-like growth factor II receptor as a multifunctional binding protein. *Nature*, **329**, 301–7.

Wallis, M. (1987) Growth-hormone receptor cloned. *Nature*, **330**, 521–2.

3 Guanine-nucleotide binding proteins as signal transducers

G-proteins as signal transducers

It has been known for several decades that cells contain special proteins that can bind guanine nucleotides (such as GTP) and which use the energy contained within GTP to regulate their conformation. By so doing, these proteins also alter their ability to interact with other proteins in an allosteric manner. These guanine-nucleotide-binding proteins (which are usually referred to as 'G' or, sometimes, 'N' proteins), therefore, represent a class of molecules which can transduce signals between separate proteins. They are found in both prokaryotes and eukaryotes and have been identified as components of a variety of 'information' systems within cells, including protein synthesis, cytoskeletal organization, visual transduction and intra-cellular messenger generation. This chapter will focus on the role of G-proteins as transducers of extracellular signals and will not consider their role in protein synthesis or in the regulation of cytoskeletal organization. Suffice it to say in this context, that several of the cofactors involved in the various stages of protein synthesis (e.g. initiation factor-2, elongation factors Tu and 1_α, and the termination protein, 'releasing factor') are GTP-binding proteins with GTPase activity, and that both the $\alpha + \beta$ subunits of the cytoskeletal protein tubulin can also bind GTP. In this case, GTP hydrolysis is essential to the control of microtubule polymerization.

A number of separate experimental observations have provided the basis for our current appreciation of the functional roles of G-proteins as signal transducers, including:

1. GTP is absolutely required for hormonal activation of membrane-bound enzymes such as adenylate cyclase.
2. This requirement for GTP can be met by hydrolysis-resistant analogues of GTP, demonstrating that GTP hydrolysis is not required for the activity of the protein to be expressed.
3. Hormones stimulate the intrinsic GTPase activity of G-proteins.
4. Binding of a hormone to its receptor promotes the exchange of bound GDP for GTP on particular G-proteins.

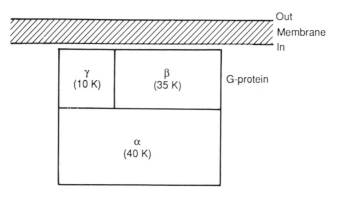

Fig. 3.1 Basic structure of G-proteins used in hormone signalling systems

Composition of G-proteins

It has become apparent within the last few years that while G-proteins represent a whole family of molecules, many of them share certain common characteristics. In particular, the G-proteins that have, so far, been directly implicated in hormone signalling are all multi-subunit proteins composed of three types of subunit, which are designated α, β and γ. These are present in equal stoichiometry in the holoprotein which is composed of one of each type of subunit monomer. The function and specificity of each G-protein is determined by the α-subunit which is unique for each type of G-protein. In contrast, the $\beta + \gamma$ subunits show much less heterogeneity and the β-subunits may be common components of all members of the G-protein family. The γ-subunits are also all very similar in structure although they may vary slightly between certain G-proteins. The variability between G-protein α-subunits is essential to their function, but even though they have structural differences in the regions involved in both receptor and effector recognition, they also share similarities in those other regions of the molecule which bind and hydrolyse GTP.

Table 3.1 lists the G-proteins that are currently believed to be involved in hormone action and provides a summary of the differences between their α-subunits.

G-protein α-subunits

α_s

The first G-protein to be characterized in detail was G_s which controls the extent of activation of adenylate cyclase by hormones. The mechanics of this process are described in Chapter 4. The stimulation of adenylate cyclase by G_s remains the prototype for the mechanism of action of other G-proteins and

Table 3.1 G-proteins: structural differences between α-subunits

G-protein	α-subunit molecular weight	Transduction system
α_s	45 000	adenylate cyclase (stimulatory)
α_{i1}	40 400	adenylate cyclase (inhibitory)
α_{i2}	40 500	adenylate cyclase ?
α_{i3}	40 500	
α_o	39 000	Unknown
$\alpha_{transducin-1}$	40 000	cGMP phosphodiesterase
$\alpha_{transducin-2}$	40 500	cGMP phosphodiesterase
α_p	40 000(?)	Phospholipase C?
α_k	41 000	Potassium channel?

most other systems seem to have adopted variations on this general theme. The α-subunit of G_s ($G_{s\alpha}$) shows some variation in its biochemical characteristics from one cell type to another but this variation in structure has not yet been linked to any fundamental difference in function. Thus, in all cells where it is present, G_s couples hormone receptors to activation of adenylate cyclase.

The structural heterogeneity of G_s was first observed by examination of the purified α-subunit following polyacrylamide gel electrophoresis under reducing conditions. The purified protein migrates as two distinct bands with apparent molecular weights of 45 K and 52 K. The relative amount of each particular component is tissue-specific but the reason for the differences is not known. In fact, it is possible that both proteins may actually be products of the same gene, since hybridization experiments have revealed that cells may contain as many as four types of α_s mRNA, which are almost identical in sequence. The slight differences probably reflect the presence of separate mRNA-processing mechanisms rather than separate genes and these give rise to each of the different products from a single primary transcript. Indeed, when the gene for the 45-K $G_{s\alpha}$ subunit was cloned and expressed in bacteria, the protein product was found to migrate with an apparent molecular weight of 52 K. This suggests that this protein may possess certain structural features which cause it to exhibit unusual behaviour on electrophoresis, and that the two forms of the protein are much more closely related than might be supposed.

α_i

The amino acid sequence of α_i has been deduced from cDNA clones obtained from both human and bovine cDNA libraries. These studies have revealed that there may be as many as three types of α_i which are differentially expressed in cells. These are all closely related to one another since they are about 90% homologous at the level of the amino acid sequence. However, a greater degree of diversity is revealed if the nucleotide sequences of the respective cDNAs are compared, suggesting that they are probably the products of at least two separate genes. It is particularly interesting that the regions of the amino acid sequence where the various forms of α_i differ comprise regions of the molecule that are envisaged to be involved in effector coupling. Thus, one speculation is that the various types of α_i may be linked to different signalling mechanisms. To date, α_i has been considered to be involved exclusively in the hormonal inhibition of adenylate cyclase (see Chapter 4), an assumption which may turn out to be an over-simplification. However, it remains to be conclusively proven that the other forms of the molecule are involved in different processes in the cell.

α_o

During the attempts to purify G_i from brain it was discovered that another G-protein is also expressed in this tissue, which shares some of the characteristics of G_i (for example, it is a substrate for ADP-ribosylation by pertussis toxin) but which is clearly a different protein. This protein is 5–10 times more abundant in brain than G_i and has the same characteristic trimeric structure with the exception that its α-subunit is slightly different, having a lower molecular weight (39 K). The genes for rat and bovine α_o have both been cloned and the amino acid sequence of the protein deduced. This has revealed that the protein has regions which are very similar to particular sites within both α_s and α_i and that it contains a binding site for guanine nucleotides. Thus, α_o is almost certainly a signal transducer but, at present, the system to which it is linked remains a mystery. It has, however, been revealed by *in vitro* reconstruction experiments that both muscarinic cholinergic and α_2-adrenergic receptors can be coupled to G_o and there is evidence that D_2-dopamine receptors may interact with this protein in anterior pituitary cells. It has also been suggested that G_o may be involved in the regulation of calcium channels by opiate agonists in neuronal cells. The protein has also been found in adrenal chromaffin cells where its abundance and distribution point to a crucial regulatory role.

α_t

Transducin is a G-protein which is localized exclusively within the photo-sensitive rod cells of the retina where it plays a role in the process of visual transduction. As such, this protein participates in a pathway which is not activated by the binding of a chemical agonist to a receptor but which is regulated by light absorption. This emphasizes the general importance of G-proteins as signal transducers and reinforces the versatility which enables them to function in a variety of different biochemical mechanisms in cells. This is not intended to imply that there is no receptor protein in the visual transduction process since this role is subserved by the photosensitive pigment rhodopsin. However, the essential difference lies in the 'agonist' which, in this case, is light itself, the signal being transmitted by an isomerization reaction within rhodopsin which then, in turn, induces transducin activation.

When the gene for the α-subunit of transducin was cloned and sequenced it was recognized that the protein exists in two forms – these are now referred to as tα_1 and tα_2 and they differ from one another in some 20% of the amino acid sequence. It is believed that these two closely-related molecules may represent the active form of 'transducin' in different cell types within the retina. Gt_1 appears to be located exclusively in rod cells while the slightly larger Gt_2 (354 amino acids compared to 350) is found only in cones. Both proteins perform the same funcion in their respective cell types, namely activation of a phosphodiesterase enzyme which specifically hydrolyses cGMP. The α-subunit of transducin-1 serves as a substrate for both pertussis and cholera toxins which ADP-ribosylate particular cysteine and arginine residues, respectively. This modification leads either to transducin activation (in the case of cholera toxin) or uncoupling from rhodopsin (pertussis toxin).

G-protein β-subunits

The availability of purified G-proteins has allowed careful comparisons to be made of the $\beta + \gamma$ subunits from different molecules, and this has revealed a remarkable similarity. In particular, it has been observed that the $\beta\gamma$ subunits of one G-protein can be functionally interchanged with those of another. For example, $\beta + \gamma$ from G_i can be used to reconstitute functional transducin if combined with α_t, or functional G_s if added to α_s. This suggests that any differences in the structure of $\beta + \gamma$ between different G-proteins are likely to be minimal.

Purified G-protein β-subunits do, however, exhibit some degree of heterogeneity. Nominally, the β-subunits are said to have a molecular weight of about 35 K, but on polyacrylamide gel electrophoresis under reducing conditions the purified 'protein' actually migrates as a doublet. Immunological evidence has been gathered which suggests that these may represent slightly different molecular forms of the protein and this evidence has now been corroborated by analysis of the amino acid sequence of cloned genes. This was first achieved for the β-subunit of transducin which was shown to contain 340 amino acids and to have a molecular weight of 37.4 K. More recently, a second β-subunit gene sequence has been deduced from adrenal tissue. This gene also codes for a protein of 340 amino acids but this molecule has a slightly different amino acid sequence from that of $G_{t\beta}$ (10% of the residues are altered). It appears, therefore, that there are two types of β-subunit in cells, and assuming that these give rise to the doublet observed on polyacrylamide gels, many cells probably express both genes. One notable exception to this is the rod cell of the retina which has only one form of β-subunit.

The functional implications of the existence of distinct types of β-subunit are not clear. Indeed, it is not known whether each of them is preferentially associated with a particular type of α (or γ) subunit(s) or whether they can both act interchangeably in different G-proteins. At present, there is little evidence to support a functional distinction, but more work will be required to resolve this issue completely.

G-protein γ-subunits

The γ-subunit of G-proteins was first positively identified as a component of transducin and then subsequently of G_s and G_i, purified from human erythrocytes. It was recognized as a low-molecular-weight component and, in the cell, it seems to remain in close association with the β-subunit during G-protein dissociation. The amino acid sequence of the γ-subunit of transducin has been determined by both genetic and direct analytical methods and it is found to contain mainly hydrophilic residues (molecular weight ~ 8.4 K). Surprisingly, antibodies raised against this molecule do not cross react with the γ-subunits of other G-proteins suggesting that there are, at least, two separate types. It is possible that one of these is confined to the rods of the retina, and that other cells express the alternative form as part of all of their G-proteins. However, antibody studies have begun to reveal evidence of divergence amongst the γ-subunits of other G-proteins, suggesting that the situation may turn out to be much more complex than this.

Originally, it was envisaged that the α-subunits of G-proteins are the all-important components responsible for fulfilling the signal transducing role of these proteins. According to this view, the $\beta + \gamma$ subunits simply serve as membrane anchors which control the availability of α for interaction with defined effector molecules. However, if, as seems likely, $\beta + \gamma$ are, themselves, structurally diverse, this implies that they may undertake more significant regulatory roles than was previously supposed. Such diversity would seem unwarranted for a simple anchorage system. One possibility that has now been raised is that the βγ complex may exert direct regulatory functions in its own right. Thus, experimental evidence has been offered that a particular type of K^+ channel in chick heart cells may be directly regulated by βγ. These experiments employed a technique known as patch clamp analysis which involves the isolation of a small 'patch' of plasma membrane on a microelectrode. Current flow through any channels present in the area of membrane can then be monitored. Using this method it has been demonstrated that purified βγ subunits from bovine brain can activate a K^+ channel which is normally regulated by muscarinic cholinergic receptors. A channel opening could not be reproduced by addition of the α-subunits of various G-proteins and the conclusion was drawn that this channel responds to the free βγ-subunits. This evidence, therefore, suggests that $\beta + \gamma$ may have direct regulatory roles in cells. However, some caution is needed as other studies have not yet verified this conclusion. In fact, similar experiments performed by other workers have produced rather different results. These studies have revealed that K^+ channels in heart cells are opened by the α-subunit of a special G-protein termed G_k. In contrast, the $\beta + \gamma$ subunits of this molecule (which are apparently identical to those of G_s) were not able to promote a channel opening. Thus, it is evident that more work must be done before a clearer picture of the role of $\beta + \gamma$ as signal transducers can be gained.

Activation of G-proteins

The topology of signalling systems in which G-proteins are involved is such that the interaction of an agonist molecule with its receptor leads to the transmission of information across a lipid membrane (in most instances). This requires that at least one of the components of the system must actually traverse the lipid bilayer and interact with other molecules located on the cytoplasmic face of the membrane. In principle, this function could be carried out by either the receptor itself or by one of the subunits of the G-protein. In practice, it is likely that all of the G-protein subunits are located on the cytoplasmic face of the membrane and that functionally significant interactions between the agonist-occupied receptor and the α-subunit of G-proteins (remember that it is the α-subunit which confers G-protein specificity) are restricted to the cytosolic domain (Fig. 3.2). Thus, it is likely to be a unique property of each individual hormone receptor which allows it to transmit information about agonist binding across the lipid bilayer (presumably via an appropriate conformational change). This information is then detected, amplified and transduced by the G-protein, which resides on the cytosolic face of the membrane. This organization is seen most clearly in the case of transducin which, as emphasized above, is slightly unusual in that it does not interact with a classical hormone receptor. Nevertheless, it does interact with rhodopsin, which is a membrane protein, and it is known that this interaction can be modulated by soluble factors which are located exclusively in the cytosol. Among the most important of these is a soluble protein, arrestin, which can inhibit the rhodopsin activation of transducin under certain circumstances. It follows then, that transducin and rhodopsin must interact at a site which is accessible to non-membrane components and which is likely to be located well below the membrane surface, i.e. in the cytosol.

The nature and significance of receptor (R)–G-protein (G) interactions has been established by studying the properties of the purified components in reconstitution experiments. It has emerged that an α-helical sequence of amino

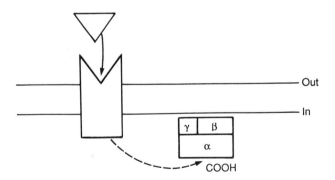

Fig. 3.2 Interactions between receptor and G-protein occurring in the cytosolic domain

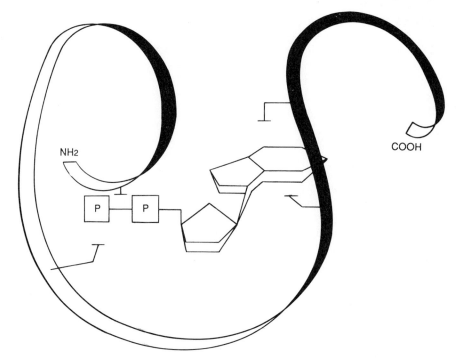

Fig. 3.3 Diagrammatic representation of the binding pocket for GDP in G-proteins

acids close to the carboxy-terminus of G-protein α-subunits is important for R–G interaction. In all G–proteins studied, this sequence contains a positively-charged residue (either arginine or lysine) which may be particularly important in this respect. For example, the defective form of α_s present in a line of S49 lymphoma cells whose α_s cannot couple to receptors has a proline for arginine replacement at this position. This seems to be the only defect in the molecule, which argues strongly for a special role for this residue in receptor G-protein recognition.

It is clear that in the unstimulated state, G-proteins contain GDP bound at a specific site within their structure. Examination of the amino acid sequences of G-protein α-subunits has revealed a marked homology between different proteins in four separate regions which are thought to adopt a three-dimensional configuration which allows the formation of the GDP-binding site. The precise coordination has been established for the GDP-binding site of elongation factor-Tu on the basis of X-ray crystallographic studies, and is likely to be similar in other G-proteins. In Ef-T$_u$, GDP is held by the formation of bonds with four specific amino acids; two of these (an aspartate and a lysine residue) interact with the β-phosphate group of GDP while two others form bonds with substituents of the guanine ring itself (Fig. 3.3). It is suggested that the conformational change necessary for activation is brought about, at least in

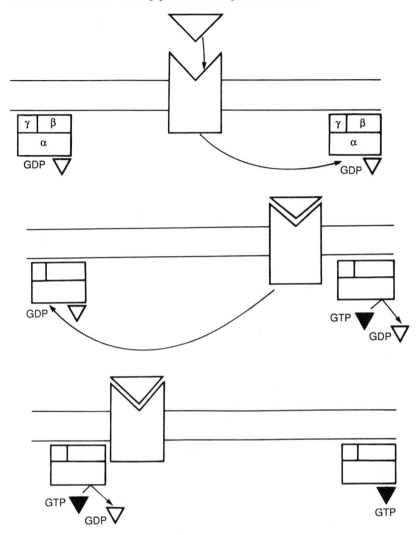

Fig. 3.4 The catalytic role of the hormone-bound receptor

part, by the spatial re-organization which occurs in order to facilitate the bonding of the extra (γ) phosphate group present in GTP.

Facilitation of the exchange of bound GDP for unliganded GTP is the primary reaction which is promoted upon interaction of a G-protein with an agonist-occupied receptor. This means that in a hormone-stimulated cell, proportionately larger numbers of G-proteins will contain bound GTP at any given time than is the case in the basal state. This, in turn, accounts for the observation that hormones (H) stimulate the intrinsic GTPase activity of G-proteins, an effect which results entirely from the increased probability that at any given time the bound ligand will be GTP rather than GDP.

Once an activated receptor has encountered and, in turn, activated a G-protein (by promoting GTP–GDP exchange) the H–R–G complex dissociates leaving the free G-protein, and allowing the H–R complex to recycle and to activate further G-proteins. This is, therefore, an important amplification step in the overall process and allows the receptor to play a catalytic role. Amplification also occurs at the next step, the interaction of G–GTP with the effector molecule (Fig. 3.4). Although G-proteins spontaneously hydrolyse GTP, the rate at which this occurs is quite slow even under receptor-stimulated conditions (turnover number is ~ 1–$2/\text{min}$). Thus, the G-protein can induce relatively prolonged effector activation before the cycle is terminated by GTP hydrolysis.

Subunit dissociation

Activation of a G-protein by a receptor proceeds by facilitation of the exchange of bound GDP for GTP as described above, and it is this binding of GTP that directly promotes dissociation of the holoprotein into its constituent subunits. It appears that the effect of dissociation is to release the α-subunit leaving the $\beta + \gamma$ subunits in close association. These latter species remain fixed in the plane of the plasma membrane and are probably peripheral membrane proteins in their native state. In contrast, the α-subunits may well be released into the soluble domain of the cell following dissociation (Fig. 3.5). Indeed, isolated α-subunits cannot interact with cellular membranes unless the membranes are first supplemented with free $\beta + \gamma$ subunits. This emphasizes the fact that the ability of α to alter the functional status of specific effector molecules relies on interactions that occur away from the plasma membrane. This may seem surprising since, in many cases, the effector molecules are, themselves, localized to the membrane fraction of the cell (e.g. adenylate cyclase) and implies that G-proteins must interact with cytoplasmic domains of these proteins to achieve their effects.

Under normal physiological conditions, GTP serves as the activating ligand promoting G-protein dissociation, but this requirement can also be fulfilled by other agents in broken cell preparations. In particular, widespread experimental use has been made of special analogues of GTP which are resistant to hydrolysis by the GTPase of the α-subunits. These include GTPγS and GppNHp which cause essentially irreversible activation of G-proteins by preventing occupation of the guanine-nucleotide binding site by GDP (Fig. 3.6); hence, they also prevent subunit re-association. These molecules have been (and continue to be) extremely valuable research tools for use in the study of G-protein function. A second powerful activator of G-protein dissociation, which can be effective in both broken cells and also, in some cases, in whole cells (e.g. liver), is (rather surprisingly!) the fluoride ion. This has been used for many years as an activator of adenylate cyclase but it is only relatively recently that it has been realized that this effect reflects G-protein activation. Indeed, the response is not unique to G_s but seems to be a characteristic of several (perhaps all) G-proteins. It has emerged that the response is not attributable

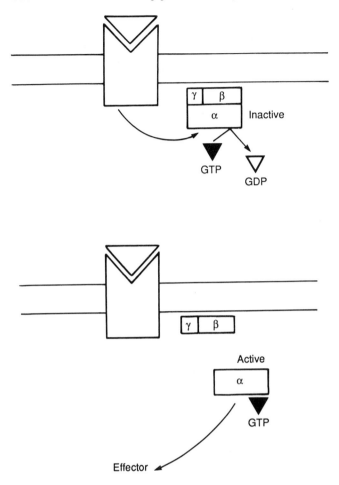

Fig. 3.5 Subunit dissociation during G-protein activation

solely to F^- but that aluminium ions (Al^{3+}) are also required. Aluminium and fluoride ions can form a range of complexes in solution, the stoichiometry of which varies according to the respective ratio of each ionic species. Titration studies point to one particular complex as the active species responsible for promoting G-protein dissociation, namely $[AlF_4]^-$. It has been argued that the ability of this ion to cause G-protein activation results from an interaction with the protein in its GDP-bound form. The interaction is such that the combined presence of GDP and $[AlF_4]^-$ at the guanine nucleotide binding site induces a conformational change in the protein which is equivalent to that promoted by GTP itself. The result is, therefore, subunit dissociation and release of the α-subunit which is then free to activate effector molecules.

It appears that subunit dissociation and subsequent re-association form a cycle which is absolutely essential for continued G-protein activation (Fig.

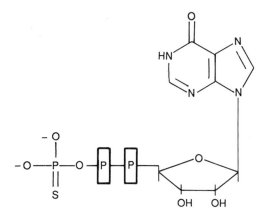

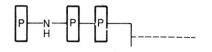

Guanosine-5'-O-(3-thiotriphosphate) : GTP-γ-S

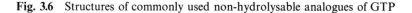

Guanylyl (βγ-imido) diphosphate : GppNHp

Fig. 3.6 Structures of commonly used non-hydrolysable analogues of GTP

3.7). Thus, once an α-subunit has hydrolysed its bound GTP to form GDP, there is an absolute requirement for re-association with $\beta + \gamma$ before the protein can be re-activated by an agonist-occupied receptor. The rate of spontaneous GTP–GDP exchange from free α-subunits is extremely slow, since the affinity for GDP is so great that it is exceeded by as much as 10^3–10^4 times in the intact cell, i.e. the affinity for GDP is micromolar or less, whereas the intracellular concentration is in the millimolar range. Thus, H–R interaction is necessary to ensure efficient exchange, and this only occurs when the protein is in its holo form.

Because G-proteins exhibit this cycle of association/dissociation, the concentration of free $\beta\gamma$ subunits in a membrane can have a direct effect on the number of free α-subunits. If, for example, the number of free $\beta\gamma$ subunits is increased for any reason, they will tend to compete with any free α-subunits and will drive the equilibrium in favour of re-association. This may be a physiologically important mechanism since in cells where two G-proteins are present the selective activation of one can lead to effective inhibition of the other. This will be especially significant if one particular G-protein is present in excess, since its activation could significantly elevate the number of $\beta\gamma$-subunits available to interact with the α-subunits of the more minor component. Indeed, this mechanism may account, at least in part, for the ability of G_i to inhibit adenylate cyclase (see Chaper 4), since many cells contain up to 10 times more G_i than G_s.

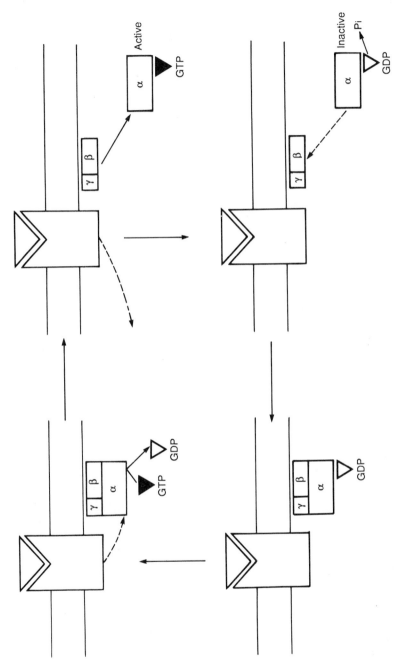

Fig. 3.7 Cycle of G-protein activation/inactivation

Bacterial toxins and G-protein function

In the mid 1970s a mechanism was discovered which has turned out to be rather more general than was originally suspected, and has been exploited to yield information about the cellular functions of G-proteins. The initial observation was that the toxic effect of diphtheria infection results from a covalent modification of elongation factor-2 (a G-protein) which effectively inhibits protein synthesis. This modification was found to be an ADP-ribosylation of the protein which occurred at a single amino acid residue. A similar mechanism was also identified for the action of the toxin of an unrelated organism, *Vibrio cholerae*, which causes unrestrained adenylate cyclase activation in cells. The normal cellular target for cholera toxin is the α-subunit of G_s, but it can also be ADP-ribosylate transducin. A specific arginine residue has been identified as the site of modification in transducin and it is possible that this is also true for α_s. The significance of the modification lies in the observation that ADP-ribosylated α_s exhibits a much reduced GTPase

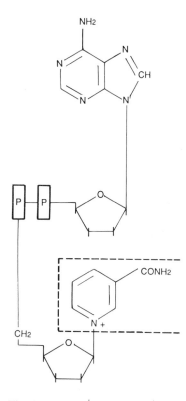

Fig. 3.8 NAD$^+$ serves as substrate for cholera and pertussis toxins which transfer ADP-ribose to proteins

activity compared to normal, which means that GTP remains bound to α_s for a very prolonged period. This has the effect of maintaining the molecule in an activated state and provides the basis for the unregulated activation of adenylate cyclase characteristic of infected cells. One experimental use to which this observation has been put depends upon the fact that the cellular substrate acts as the source of the ADP-ribose is NAD^+. If, therefore, NAD^+ is supplied in a radioactive form (e.g. as $^{32}P\text{-}NAD^+$) cholera toxin will catalyse the covalent transfer of ^{32}P-labelled ADP-ribose to its substrate G-protein (Fig. 3.8). This has been useful as a means to identify and characterize the different forms of α_s present in cells.

It has been established that although α_s is the substrate for cholera toxin, a second protein also plays an obligatory role in the catalytic mechanism of the reaction catalysed by the toxin. This molecule has been termed ADP-ribosylation factor (ARF) and is, itself, a guanine nucleotide binding protein. It is believed that a GTP-bound form of ARF participates in the cholera toxin mediated ADP-ribosylation of α_s and that α_s has to contain bound GDP for the reaction to occur. The $\beta + \gamma$ subunits of G_s do not, however, appear to be necessary, since purified α_s can serve as a substrate for cholera toxin in the presence of ARF. It is tempting to speculate that ARF also exerts some physiological control over α_s in the intact cell but the evidence to support this idea has not yet been forthcoming.

Another bacterial toxin which has also proved useful for G-protein identification, is produced by the organism responsible for whooping cough, *Bordetella pertussis*. This toxin was first identified by virtue of its ability to antagonize the inhibitory effects of catecholamines on the rate of insulin secretion from the B-cells of islets of Langerhans. It was, therefore, originally termed 'islet activating protein'. However, since that time its effects have also been observed in a wide range of other cells and it has now assumed the rather less definitive title of 'pertussis toxin'. Like cholera toxin, pertussis toxin is also an ADP-ribosyltransferase but its substrate specificity is somewhat more general. The α-subunit of G_i is a good substrate for the toxin and becomes ADP-ribosylated on a cysteine residue close to the carboxy-terminus. Unlike the situation with cholera toxin, ARF is not required for ADP-ribosylation of α_i, and purified α_i does not serve as a substrate for pertussis toxin. This accords with the overall effect of pertussis toxin on the protein which is to prevent dissociation and activation of G_i. Therefore, ADP-ribosylation maintains G_i in its holo-form and prevents release of free α_i. This is opposite to the effect of cholera toxin on G_s, where ADP-ribosylation leads to an increase in the availability of free α_s. Pertussis toxin also functionally uncouples G_i from hormone occupied receptors, thereby inhibiting the interaction of such receptors with G_i.

In addition to G_i, a number of other G-proteins are also substrates for pertussis toxin. These include transducin and G_o among the better characterized molecules and several other G-proteins from both brain and peripheral tissues which are less well characterized and whose functions have not been firmly established.

ras-proteins

In recent years, a group of proteins has been identified in cells which are smaller than the well-characterized G-protein α-subunits, but which show some homologies with these proteins. These molecules are termed *ras* proteins, and they have within their amino acid sequences, regions which fold together to form a GTP-binding site similar to that of other G-proteins. Furthermore, they have a measurable GTPase activity and are believed to be localized at the inner leaflet of the plasma membrane, which are all characteristics of other G-proteins. The proteins which comprise the *ras* family are all about 21 K in molecular weight, but there is no evidence that they interact with the β or γ subunits of other G-proteins or that they have their own equivalent regulatory subunits.

The importance of *ras* proteins lies in the fact that they are associated with the malignant transformation of cells. Thus, certain retroviruses code for *ras*-related proteins (e.g. Harvey Rous sarcoma virus) and it is thought that expression of these variant molecules may be related to the transforming potential of such viruses. This concept has led to the idea that the normal cellular genes from which the viral genes have arisen (proto-oncogenes) code for proteins that control some aspect of cell growth or differentiation and that expression of the viral version leads to a breakdown of the system which normally regulates these processes. Thus, *ras* proteins are seen as potential intermediaries in a growth-regulating transduction system. So far, however, no receptor has been identified for which *ras* proteins could serve as transducers and it may prove to be the case that they are not true 'transducers' in the manner of G_s, G_i etc., but that they are direct regulators of particular signalling systems.

Each member of the *ras* protein family (which numbers as many as 10 different proteins from organisms ranging from yeast to man) has a cysteine residue near to the carboxy-terminus which is acylated by covalent attachment of a long-chain fatty acid (palmitic acid). This appears to be responsible for anchoring the protein to the plasma membrane. Since *ras* proteins are located at the plasma membrane, it has been assumed that their targets are also to be found in this compartment of the cell. Some evidence (see Chapter 5) suggests that they might be capable of regulating phospholipase C activity and that they thereby control inositol phosphate and diacylglycerol production. This would not be entirely inappropriate to their perceived role as regulators of cell growth, since several (although by no means all) growth promoters can initiate phospholipase C activation. However, it seems more likely that another, more fundamental, system also exists which is controlled by *ras* proteins and that phospholipase C activation is a secondary response which occurs in some cells in response to *ras* protein expression.

In view of the failure to detect a receptor to which *ras* may be linked, it has been suggested that these proteins could be regulated in an indirect manner, perhaps by growth factor receptors. One possibility, in this context, is that they might serve as endogenous substrates for the tyrosine kinase enzymes which form part of some growth factor receptors, such as those for insulin and epidermal growth factor (EGF) (see Chapter 6). Indeed, *ras* has been shown to act as a substrate for the EGF receptor kinase and indirect evidence suggests that some of the cellular actions of insulin can be inhibited if cells are micro-injected with anti-*ras* antibodies. This would be consistent with the involvement of *ras* in the mechanism of action of insulin.

It is important that the specific effector molecule(s) with which *ras* proteins interact are identified in cells. Only then will a firmer understanding of their role begin to emerge, but already it is evident that this group of G-proteins probably subserve vital functions in the cell, which are likely to relate to the coordination of cell growth and differentiation.

G-proteins and disease

The fact that G-proteins are such important components of hormone signalling mechanisms and that they are so widespread in their cellular distribution suggests that they are unlikely to readily tolerate structural changes such as those that might result from genetic mutation. Indeed, it is probable that many G-protein mutations would be so detrimental to cell function that they would be lethal to the organism. However, recent evidence suggests that certain disease states may be associated with defects in G-protein function. One such example is the condition known as pseudohypoparathyroidism in which the clinical symptoms are consistent with the patients having a reduced level of circulating parathyroid hormone (i.e. they have low calcium and high phosphorus concentrations in the circulation). However, determination of the circulating hormone level reveals that it is not deficient in such patients but that the problem lies in the target tissues, which fail to respond to the hormone in a normal manner. Furthermore, in some patients the syndrome is also associated with resistance to a number of other hormones suggesting that the actual defect lies at a fundamental level in the hormone signalling process itself. When the cells of such patients were assayed for G_s activity (by reconstitution into an adenylate cylase system) it was observed that their G_s failed to fully couple hormone receptors to the enzyme. The results suggested that the cells contained only 50% of the normal level of this G-protein. It therefore seems likely that these patients carry one defective gene for G_s and that their failure to express a full complement of the protein leads to the functional impairment of cellular responses which is characteristic of the disease.

More recently, alterations in G_s have also been implicated in a different disease state which is characterized by an increased synthesis of cAMP rather than by a decrease, as in the case of pseudohypoparathyroidism. The observations have arisen out of a study of one class of pituitary tumours (adenoma) which has shown that the tumour cells contain a form of G_s which is much more active than usual. These cells are therefore in a constantly activated state and have high levels of cAMP as a consequence. One surprising observation is that this 'activated G_s' is much less susceptible to ADP-ribosylation by cholera toxin than normal. The reasons for this have not yet been established, but one obvious possibility is that the molecule has already become ADP-ribosylated in the cell by an endogenous enzyme which is normally present at only very low activity. It is possible that the production of this (hypothetical) enzyme is enhanced in the tumour cells such that all of the G_s molecules become modified and are thereby converted into an irreversibly activated state. An alternative possibility is that the gene for G_s has mutated in these cells such that a 'super-active' molecule is produced which is not subject to the normal physiological regulatory mechanisms.

It is not clear whether altered G_s is a cause of tumour production in pituitary adenomas, or whether it is a secondary effect which results from cell transformation. However, it seems significant that the *ras* oncogenes are also G-proteins which, when activated, are associated with the transformed state. Further analysis of the precise nature of the modification of G_s in these adenoma cells should provide new insights into the role and regulation of this molecule in normal and transformed cells.

Another G-protein which has been implicated in human disease is the rod-cell protein transducin which controls the activity of a cGMP-phosphodiesterase and plays an important role in the visual transduction process. A condition known as human retinitis pigmentosa has been identified which is associated with progressive retinal degeneration and is preceded by the maintenance of an unusually high concentration of cGMP in the rod cells of the retina (at least in animal models). Speculation suggests that the high levels of cGMP may reflect a defect in the activity of transducin-activated phosphodiesterase and raise the possibility that transducin itself may be altered in diseased cells.

A third G-protein, G_i, has now also been implicated in the development of a disease state since it has been reported that in an animal model of type-I diabetes mellitus, there is a reduction in the ability of G_i to mediate hormonal inhibition of adenylate cyclase. This observation apparently reflects a real decrease in the amount of the active α-subunit of the protein in the livers of diabetic animals, as judged by reduced immunoreactivity against an anti$G_{\alpha i}$ antibody. The low level of G_i can be restored towards normal by insulin treatment which suggests that insulin may be an important regulator of the expression of this protein, at least in liver. A reduction in G_i may be directly related to some of the metabolic complications associated with diabetes and could cause a decrease in tissue sensitivity to certain hormones. It will be of interest to determine whether human diabetic patients also have lower than normal levels of G_i in insulin-sensitive tissues, and, if so, how effectively the level is restored by insulin therapy.

Cholera

Another disease which has already been alluded to above, cholera, may also be related to a G-protein dysfunction, although, in this case, it is not a genetically determined structural defect which causes the disease, but rather, it reflects a toxin-induced modification of the protein. The symptoms of cholera include severe dehydration caused by excessive water loss due to diarrhoea; the causitive agent is the bacterium *Vibrio cholerae*. This organism secretes an active toxin which has ADP-ribosyltransferase activity and catalyses the covalent modification of the adenylate cyclase stimulatory protein, G_s. This has the effect of fully activating G_s and leads to large-scale increases in cAMP in target cells. Many cells are susceptible to cholera toxin *in vitro* which reflects the fact that the 'receptor' to which cholera toxin binds is widespread. This receptor is a ganglioside molecule (GM_1) containing sugar residues linked to the lipid backbone of ceramide. Cholera toxin interacts with GM_1 via its B-subunits (of which there are five) and this interaction leads to the formation of a pore in the plasma membrane through which the catalytically active A-subunit is extruded into the cytosol. Once inside the cell, the toxin ADP-ribosylates G_s and thereby activates adenylate cyclase.

The reason that the physical manifestations of cholera infection are restricted to the gut relates to the fact that the producer organism specifically colonizes regions of the small intestine, therefore resulting in the preferential exposure of the cells of the intestinal mucosa to the toxin. The excessive water loss induced by cholera toxin results from an osmotic imbalance which develops across the intestinal mucosa as a result of the elevated cAMP levels. A rise in cAMP in the crypt cells promotes the secretion of chloride into the gut and also inhibits Na^+ uptake from the lumen. This has the effect of creating a hyperosmolar environment and facilitates water movement into the intestinal lumen, which then causes the characteristic diarrhoea.

Whooping Cough

The disease whooping cough results from infection with the bacterium *Bordetella pertussis* which produces a toxin (pertussis toxin) capable of modification of G_i such that inhibitory hormones are unable to attenuate the activity of adenylate cyclase. The organism responsible for this disease may have access to more tissues than that producing cholera toxin which implies that pertusis toxin can potentially interact functionally with a number of different types of cell. This means that infection can be associated with a range of symptoms not all of which are related to the respiratory tract. One of the major effects of *Bordetella* infection is to sensitize the host to histamine. This agent normally causes vasodilation and a drop in blood pressure in animals, and its effects are controlled by a homeostatic reflex mechanism which induces catecholamine release and compensatory vasoconstriction. In pertussis toxin-treated animals, the catecholamine effects are blocked since they are normally mediated by α_2-adrenoceptors which are coupled to G_i. This means that a 'super-sensitivity' to histamine develops and the uncontrolled vasodilation can be fatal.

A second symptom of patients infected with *Bordetella pertussis* is hypoglycaemia. This probably results from the failure of an inhibitory control mechanism which normally regulates the rate of insulin secretion, and is mediated by α_2-adrenoceptors on the B-cells of pancreatic islets. The inactivation of this mechanism by toxin-catalysed modification of G_i leads to abnormally elevated rates of insulin secretion which then precipitates the hypoglycaemia.

ras oncogenes

The possibility that a special group of G-proteins, the *ras* proteins, may be involved in the development of some types of cancer has come to the fore in recent years, since it was realized that variants of these proteins are encoded by genes carried by some oncogenic viruses. At least three *ras* genes are present in mammalian cells and they all code for proteins of about 21 K in molecular weight which have intrinsic GTPase activity. Over-expression of these proteins can induce transformation of cells in culture as can the synthesis of a mutant variant which has only a single amino acid substitution relative to the wild-type protein. Indeed, such mutants have been isolated from lung and bladder carcinoma cells which strongly implicates them in the development of some human tumours. Normally, *ras* proteins hydrolyse GTP by virtue of their intrinsic catalytic activity and the rate of this reaction is comparable to that of other cellular G-proteins (i.e. they have a low turnover number). The oncogenic variants of *ras*, which have been studied so far, all have a still slower rate of GTP hydrolysis than normal, and in some proteins the activity is essentially absent. This means that if the proteins are activated by GTP (as is the case for other G-proteins) they will be maintained in an artificially activated state by these mutations, and could induce uncontrolled activation of the specific cellular process which they regulate. At present, the nature of these processes remains a mystery but it seems very likely that they must be systems which are directly involved in the regulation of cell growth. Furthermore, the current evidence points strongly to the conclusion that G-protein (*ras* protein) malfunction is probably a central component in the development of certain kinds of cancer.

Further reading

G-proteins – general

Bourne, H.R. (1986). One molecular machine can transduce diverse signals. *Nature*, **321**, 814–16.

Bourne, H.R., Masters, S.B. and Sullivan, K.A. (1987). Mammalian G proteins: structure and function. *Biochem. Soc. Trans.*, **15**, 35–8.

Casey, P. (1988). G-protein involvement in receptor–effector coupling. *J. Biol. Chem.*, **263**, 2577–80.

Chabre, M. (1987). The G-protein connection: is it in the membrane or the cytoplasm? *Trends Biochem. Sci.*, **12**, 213–15.

Chabre, M (1987). Receptor–G protein precoupling: neither proven nor needed. *Trends Neurosci.*, **10**, 335–6.

Drummond, A.H. (1988). Lithium affects G-protein receptor coupling. *Nature*, **331**, 338.

Eide, B., Gierschik, P., Milligan, G. *et al.* (1987). GTP-binding proteins in brain and neutrophil are tethered to the plasma membrane via their amino termini. *Biochem. Biophys. Res. Commun.*, **148**, 1398–405.

Gilman, A.G. (1984). G proteins and dual control of adenylate cyclase. *Cell*, **36**, 577–9.

Gilman, A.G. (1987). G-proteins: transducers of receptor-generated signals. *Ann. Rev. Biochem.*, **56** 615–49.

Hughes, S.M. (1983). Are guanine nucleotide binding proteins a distinct class of regulatory proteins? *FEBS Lett.*, **164**, 1–8.

Litosch, I. (1987). Regulatory GTP-binding proteins: emerging concepts on their role in cell function. *Life Sci.*, **41**, 251–8.

Lochrie, M.A. (1988). G-protein multiplicity in eukaryotic signal transduction systems. *Biochemistry*, **27**, 4956–65.

Lynch, C.J., Morbach, L., Blackmore, P.F. and Exton, J.H. (1986). α-subunits of N_s are released from the plasma membrane following cholera toxin activation. *FEBS Lett.*, **200**, 333–6.

Milligan, G., Gierschik, P. and Spiegel, A.M. (1987). The use of specific antibodies to identify and quantify guanine nucleotide-binding proteins. *Biochem. Soc. Trans.*, **15**, 42–5.

Spiegel, A.M. (1987). Signal transduction by guanine nucleotide binding proteins. *Mol. Cell. Endocrinol.*, **49**, 1–16.

Spiegel, A.M., Gierschink, P., Levine, M.A. and Downs, R.W. (1985). Clinical implications of guanine nucleotide-binding proteins as a receptor–effector couplers. *N. Engl. J. Med.*, **312**, 26–33.

Stryer, L. and Bourne, H.R. (1986). G proteins: a family of signal transducers. *Ann. Rev. Cell Biol.*, **2**, 391–419.

Sullivan, K.A., Miller, R.T., Masters, S.B. *et al.* (1987). Identification of receptor contact site involved in receptor–G-protein coupling. *Nature*, **330**, 758–60.

G-protein α-subunits

Beals, C.R., Wilson, C.B. and Perlmutter, R.M. (1987). A small multigene family encodes G_i signal-transduction proteins. *Proc. Natl. Acad. Sci. USA*, **84**, 7886–90.

Bray, P., Carter, A., Simons, C. *et al.* (1986). Human cDNA clones for four species of $G_{\alpha s}$ signal transduction protein. *Proc. Natl. Acad. Sci. USA*, **83**, 8893–7.

Bray, P., Carter, A., Guo, V. *et al.* (1987). Human cDNA clones for an α-subunit of G_i signal-transduction protein. *Proc. Natl. Acad. Sci. USA*, **84**, 5115–19.

Dickey, B.F., Pyun, H.Y., Williamson, K.C. and Navarro, J. (1987). Identification and purification of a novel G-protein from neutrophils. *FEBS Lett.*, **219**, 289–92.

Didsbury, J.R. and Snyderman, R. (1987). Molecule cloning of a new human G protein – evidence for two $G_{i\alpha}$-like protein families. *FEBS Lett.*, **219**, 259–63.

Didsbury, J.R., Ho, Y-S. and Snyderman, R. (1987). Human G_i protein α-subunit: deduction of amino acid structure from a cloned cDNA. *FEBS Lett.*, **211**, 160–4.

Goldsmith, P., Gierschik, P., Milligan, G. *et al.* (1987). Antibodies directed against synthetic peptides distinguish between GTP-binding proteins in neutrophil and brain. *J. Biol. Chem.*, **262**, 14683–8.

Graziano, M.P., Freissmuth, M. and Gilman, A. (1989). Expression of $G_{s\alpha}$ in *Escherichia coli*. purification and properties of two forms of the protein. *J. Biol. Chem.*, **264**, 409–18.

Hanley, M.R. and Jackson, T. (1987). Transformer and transducer. *Nature*, **328**, 668–9.

Iyengar, R., Rich, K.A., Herberg, J.T. *et al.* (1987). Identification of a new GTP-binding protein. *J. Biol. Chem.*, **262**, 9239–45.

Katada, T., Oinuma, M., Kusakabe, K. and Ui, M. (1987). A new GTP-binding protein in brain tissues serving as the specific substrate of islet-activating protein, pertussis toxin. *FEBS Lett.*, **213**, 353–8.

Mattera, R., Yatani, A., Kirsh, G.E. *et al.* (1989). Recombinant α_i-3 subunit of activates G_k-gated K^+ channels. *J. Biol. Chem.*, **264**, 465–71.

Murphy, P.M., Eide, B., Goldsmith, P. *et al.* (1987). Detection of multiple forms of $G_{i\alpha}$ in HL60 cells. *FEBS Lett.*, **221**, 81–6.

Robishaw, J.D., Smigel, M.D. and Gilman, A.G. (1986). Molecular basis for two forms of the G-protein that stimulates adenylate cyclase. *J. Biol. Chem.*, **261**, 9587–90.

Tanube, T., Nukada, T., Nishikawa, Y. *et al.* (1985). Primary structure of the α-subunit of transducin and its relationship to *ras* proteins. *Nature*, **315**, 242–5.

Van Meurs, K.P., Angus, C.W., Lavu, S. *et al.* (1987). Deduced amino acid sequence of bovine retinal $G_{o\alpha}$: similarities to other guanine nucleotide-binding proteins. *Proc. Natl. Acad. Sci. USA*, **84**, 3107–11.

G-protein $\beta + \gamma$ subunits

Bokoch, G.M. (1987). The presence of free β/γ subunits in human neutrophils results in suppression of adenylate cyclase activity. *J. Biol. Chem.*, **262**, 589–94.

Gao, B., Gilman, A.G. and Robishaw, J.D. (1987). A second form of the β-subunit of signal transducing G proteins. *Proc. Natl. Acad. Sci. USA*, **84**, 6122–5.

Hildebrandt, J.D., Codina, J., Risinger, R. and Birnbaumer, L. (1984). Identification of a γ-subunit associated with the adenylate cyclase regulatory proteins N_s and N_i. *J. Biol. Chem.*, **259**, 2039–42.

Rosenthal, W., Koesling, D., Rudolph, U. *et al.* (1986). Identification and characterization of the 35-kDa β-subunit of guanine-nucleotide-binding proteins by an antiserum raised against transducin. *Eur. J. Biochem.*, **158**, 255–63.

G-protein function

Andrade, R., Malenka, R.C and Nicoll, R.A (1986). A G protein couples serotonin and $GABA_\beta$ receptors to the same channels in hippocampus. *Science*, **234**, 1261–5.

Arad, H., Rosenbusch, J.P. and Levitzki, A. (1984). Stimulatory GTP regulatory unit N_s and the catalytic unit of adenylate cyclase are tightly associated: Mechanistic consequences. *Proc. Natl. Acad. Sci. USA*, **81**, 6579–83.

Barrowman, M.M., Cockcroft, S. and Gomperts, B.D. (1986). Two roles for guanine nucleotides in the stimulus–secretion sequence of neutrophils. *Nature*, **319**, 504–7.

Bourne, H. (1987). 'Wrong' subunit regulates cardiac potassium channels. *Nature*, **325**, 296–7.

Bourne, H.R. (1987). Discovery of a new oncogene in pituitary tumours. *Nature*, **330**, 517–18.

Codina, J., Grenet, D., Yatani, A. *et al.* (1987). Hormonal regulation of pituitary GH_3 cell K^+ channels by G_k is mediated by its α-subunit. *FEBS Lett.*, **216**, 105–6.

Codina, J., Yatani, A., Grenet, D. *et al.* (1987). The α-subunit of the GTP binding protein G_k opens atrial potassium channels. *Science*, **236**, 442–5.

Dunlap, K., Holz, G.G. and Rane, S.G. (1987). G-proteins as regulators of ion channel function. *Trends Neurosci.*, **10**, 241–3.

Fain, J.N. (1988). Evidence for the involvement of G-proteins in activation of phospholipases by hormones. *FASEB J.*, **2**, 2569–74.

Gawler, D., Milligan, G., Spiegel, A.M. *et al.* (1987). Abolition of the expression of inhibitory guanine nucleotide regulatory protein G_i activity in diabetes. *Nature*, **327**, 229–32.

Gibbs, J.B., Sigal, S. and Scolnick, E.M. (1985). Biochemical properties of normal and oncogenic *ras* p21. *Trends Biochem. Sci.*, **10**, 350–3.

Gomperts, B.D. (1983). Involvement of guanine nucleotide-binding protein in the gating of Ca^{2+} by receptors. *Nature*, **306**, 64–6.

Haga, K., Haga, T., Ichiyama, A. *et al.* (1985). Functional reconstitution of purified muscarinic receptors and inhibitory guanine nucleotide regulatory protein. *Nature*, **316**, 731–3.

Haslam, R.J. and Davidson, M.M.L. (1984). Guanine nucleotides decrease the free Ca^{2+} required for secretion of serotonin from permeabilised blood platelets. Evidence of a role for a GTP-binding protein in platelet activation. *FEBS Lett.*, **174**, 90–5.

Holz IV, G.G., Rane, S.G. and Dunlap, K. (1986). GTP-binding proteins mediate transmitter inhibition of voltage-dependent calcium channels. *Nature*, **319**, 670–2.

Knight, D.E. and Baker, P.F. (1985). Guanine nucleotides and Ca-dependent exocytosis. *FEBS Lett*, **189**, 345–9.

Logothetis, D.E., Kurachi, Y., Galper, J. *et al.* (1987). The $\beta\gamma$-subunits of GTP-binding proteins activate the muscarinic K^+ channel in heart. *Nature*, **325**, 321–5.

Pennington, S.R. (1987). G-proteins and diabetes. *Nature*, **327**, 188–9.

Plaffinger, P.J., Martin, J.M., Hunter, D.D. *et al.* (1985). GTP-binding proteins couple cardiac muscarinic receptors to a K^+ channel. *Nature*, **317**, 536–8.

Sasaki, K. and Sato, M (1987). A single GTP-binding protein regulates K^+ channels coupled with dopamine, histamine and acetylcholine receptors. *Nature*, **325**, 259–62.

Scott, R.H. and Dolphin, A.C. (1987). Activation of a G-protein promotes agonist responses to calcium channel ligands. *Nature*, **330**, 760–2.

Trahey, M. and McCormick, F. (1987). A cytoplasmic protein stimulates normal N-*ras* p21 GTPase, but does not affect oncogenic mutants. *Science*, **238**, 542–5.

G-protein ADP-ribosylation

Cote, T.E., Frey, E.A. and Sekura, R.D. (1984). Altered activity of the inhibitory guanyl nucleotide-binding component (N_i) induced by pertussis toxin. *J. Biol. Chem.*, **259**, 8693–8.

Gaal, J.C. and Pearson, C.K. (1986). Covalent modification of proteins by ADP-ribosylation. *Trends Biochem. Sci*, **11**, 171–5.

Holmgren, J. (1981). Actions of cholera toxin and the prevention and treatment of cholera. *Nature*, **292**, 413–17.

Jacquemin, C., Thibout, H., Lambert, B. *et al.* (1986). Endogenous ADP-ribosylation of G_s subunit and autonomous regulation of adenylate cyclase. *Nature*, **323**, 182–4.

Lo, W.W.Y. and Hughes, J. (1987). Pertussis toxin distinguishes between muscarinic receptor-mediated inhibition of adenylate cyclase and stimulation of phosphoinositide hydrolysis in Flow 9000 cells. *FEBS Lett.*, **220**, 155–8.

Malbon, C.C., Rapiejko, P.J. and Garcia-Sainz, J.A. (1984). Pertussis toxin catalyses the ADP-ribosylation of two distinct peptides, 40 and 41 kDa, in rat fat cell membranes. *FEBS Lett.*, **176**, 301–6.

Sekura, R.D., Fish, F., Manclark, C.R. *et al.* (1983). Pertussis toxin-affinity purification of a new ADP-ribosyltransferase. *J. Biol. Chem.*, **258**, 14647–51.

Ueda, K. and Hayaishi, O. (1985). ADP-ribosylation. *Ann. Rev. Biochem.*, **54**, 73–100.

Wregget, K.A. (1986). Bacterial toxins and the role of ADP-ribosylation. *J. Receptor Res.*, **6**, 95–126.

4 Cyclic nucleotides as second messengers

Introduction

The importance of cAMP as a second messenger involved in the mediation of hormonal responses has been recognized ever since its discovery by Dr Earl Sutherland more than 30 years ago. It was first observed as a heat-stable factor which could be generated by treatment of a particulate subcellular fraction obtained from rat liver with adrenaline. Its production required the presence of ATP and the factor could induce activation of the cytosolic enzyme glycogen phosphorylase. Subsequent demonstration that the active principle was cAMP established this molecule as the first 'second messenger' to be identified (Fig. 4.1). These discoveries were followed by the demonstration that most eukaryotic cells possess a plasma membrane-associated enzyme, adenylate cyclase, which can produce cAMP from ATP and that the activity of this enzyme can be regulated by various hormones. These observations formed the basis of the second messenger hypothesis formulated by Sutherland, which suggested that hormones interact with specific receptor sites at the surface of the cell, and that this mediates an increase in adenylate cyclase activity, thereby increasing the rate of synthesis of cAMP. By this mechanism an extracellular hormonal signal could be transduced across the plasma membrane to produce an intracellular response. The original mechanisms which Sutherland proposed in order to explain how this system might work were based on the concept of allostery, and involved a 'conformational' change in the system upon agonist binding. He envisaged that hormones could directly interact with the catalytic moiety of adenylate cyclase, leading to its activation. However, subsequent studies in several tissues revealed that the system is a multi-component one and that hormones interact not with allosteric sites on the catalytic unit, but with separate, specific receptor proteins. In addition, a third essential component has also been identified. It became evident during the early 1970s that GTP was necessary for hormonal activation of adenylate cyclase. This fact had previously been largely overlooked because the ATP used in assays of the enzyme often contained sufficient GTP as a contaminant to fulfil this requirement. However, it is now clear that the requirement for GTP reflects the presence of a transducer protein which carries a signal between the activated hormone receptor and the catalytic subunit of the enzyme. Two such proteins have been implicated in

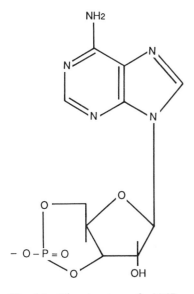

NH2

Fig. 4.1 The structure of cAMP

regulation of adenylate cyclase activity: one controls hormonal activation of the enzyme ('G_s'/N_s) while the second (G_i/N_i) mediates the effects of hormones and drugs which inhibit adenylate cyclase activity. These are distinct molecules although both have a heterotrimeric structure and undergo dissociation in response to agonist binding at appropriate receptors. Hormone-sensitive adenylate cyclase has been found in many different cell types and a wide range of physiological agonists have now been identified which control the activity of the enzyme. These range from low-molecular-weight compounds such as noradrenaline and dopamine, to large proteins such as ACTH and TSH.

Hormone receptors

It is not yet possible to provide a complete profile of the structural features of hormone receptors which are coupled to adenylate cyclase, but the discussion in Chapter 2 gives an overview of the role and regulation of receptors in general. The best studied of the cyclase-coupled receptors is the β-adrenergic receptor which has recently been purified to homogeneity. Detailed studies of the structure and function of this receptor are now in progress and should provide new insights into the molecular details of the process of adenylate cyclase activation. The β-adrenergic receptor may then serve as a prototype for other receptors that are linked to adenylate cyclase, since the basic features of all such receptors are likely to be very similar. Indeed, it seems probable that all receptors which cause adenylate cyclase activation must share similar

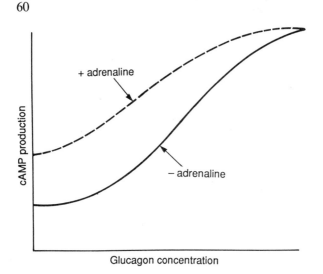

Fig. 4.2 Effects of glucagon and adrenaline on cAMP levels in liver cells

structural features which allow them to recognize, interact with and activate G_s. This is supported by data which show that in reconstitution experiments, hormone receptors from one tissue can completely activate adenylate cyclase in an unrelated tissue taken from a different species of animal. For example, glucagon receptors obtained from rat liver cells can mediate activation of the adenylate cyclase present in a human erythroleukaemia cell line. Furthermore, it appears that in any given tissue all receptors are coupled to the same pool of adenylate cyclase molecules. Hence, stimulation of rat hepatocytes with a high concentration of glucagon produces a rise in cAMP which cannot be further enhanced by addition of adrenaline, suggesting that no further activation of adenylate cyclase occurs under these conditions. Submaximal concentrations of the two hormones do, however, produce additive effects (see Fig. 4.2). Thus, receptors with different hormone specificities are able to activate the same population of adenylate cyclase molecules.

Control of adenylate cyclase activity by guanine-nucleotides

As described in detail in Chapter 3, there are two transducer proteins which control the activity of adenylate cyclase. These are both guanine-nucleotide-binding ('G') proteins and are designated G_s (stimulatory) and G_i (inhibitory).

Early observations which indicated a requirement for guanine-nucleotides in cyclase activation were interpreted in terms of an amplificaion role for these molecules but later studies revealed that they are true requirements of the activation process. Typically, concentrations of GTP in the range $10^{-8} - 10^{-4}$ M are necessary in order to facilitate hormonal regulation of the enzyme, and the effect is specific for the triphosphorylated form. Indeed, GDP and its nonphosphorylatable analogue guanosine 5-[β-thio]diphosphate (GDP-β-S) are competitive inhibitors of the ability of GTP to promote cyclase activation. Conversely, non-hydrolysable analogues of GTP (e.g. GTP-γ-S; GppNHp) can substitute for native GTP as activators of adenylate cyclase and, in some cases, they can cause cyclase activation in the complete absence of hormone. These data suggest that, unlike many reactions which require ATP, the role of GTP is not one in which the hydrolysis of the terminal phosphate is required as part of the activation mechanism. In fact, one feature of the activation of the enzyme by GTP-γ-S is that the stimulation is persistent. This indicates that when GTP hydrolysis is prevented, the enzyme is maintained in an active state and suggests that hydrolysis of the γ-phosphate of GTP is probably involved in terminating the stimulatory process. Indeed, labelling of G-proteins with ^{3}H-GTP is accompanied by release of ^{3}H-GDP which is consistent with a functional role for hydrolysis of the γ-phosphate. Interestingly, the rate of GDP release from its binding sites on guanine-nucleotide-binding proteins is directly stimulated by the presence of hormones which activate adenylate cyclase. This observation suggests that one of the mechanisms by which hormones cause enzyme acvtivation is by increasing the rate of dissociation of bound GDP, thereby facilitating the binding of GTP. A second feature of the hormone activated system is that the rate of GTP hydrolysis is also increased. At first sight this seems to be at odds with the idea that GTP is the active species responsible for stimulation of adenylate cyclase. However, it can be seen that the overall effect of these two responses will be to increase the rate of cycling of GTP through the system and, since the turnover number of the GTPase activity is quite low (approximately 1 per second), the net effect will be to increase the number of G-proteins which contain bound GTP (and are therefore active) at any given time. Despite these considerations it has been estimated that even under conditions of maximal receptor occupancy, only about 30% of the available G_s is in an active state at any time, which presumably reflects the increased GTPase activity under these conditions. The GTPase is important for switching off the enzyme activation, and the relative rates of GTP binding and hydrolysis are regulated so as to allow efficient activation and termination of the hormone signal, as appropriate. The importance of this can be readily seen under conditions where the hormone-stimulated GTPase activity is inhibited. This is the case, for example, in cells treated with cholera toxin, which induces irreversible activation of adenylate cyclase. The consequent inability to reduce intracellular cAMP levels *in vivo* represents the major underlying factor which results in water loss via the gut and produces the often fatal symptoms of the disease (see Chapter 3).

Structure and function of G_s

The transducer protein, G_s, responsible for mediating the effects of GTP on adenylate cyclase was first identified using a photoaffinity-labelled GTP analogue which bound to several proteins in avian erythrocyte membranes.

One of these had a molecular weight of 45 K and was closely associated with the catalytic activity of adenylate cyclase upon fractionation of the proteins. A labelled protein of similar molecular weight was also detected in membrane fractions from several other cell types treated with cholera toxin in the presence of ^{32}P-NAD$^+$. As outlined above, this toxin causes irreversible activation of adenylate cyclase by stabilizing the GTP-activated form of G$_s$. This effect results from covalent modification of G$_s$ by transfer of ADP-ribose from NAD$^+$ to a specific amino acid residue (possibly arginine) in the protein. Thus, cholera toxin is a protein which possesses ADP-ribosyltransferase activity and uses NAD$^+$ as substrate. The demonstration that this toxin could modify a single protein in membrane preparations isolated from cells having a functional hormone-sensitive adenylate cyclase system was then good evidence that the protein represented at least one part of G$_s$. This evidence was supported by the isolation of a mutant strain of lymphoma cells which, unlike the parental S49 cells, failed to respond to hormones by any increase in intracellular cAMP (S49 cyc$^-$ cells). They were shown to possess receptors for a variety of hormones which would normally be expected to couple to adenylate cyclase and to contain a functional catalytic subunit. Membrane proteins from these cells could not, however, be labelled with cholera toxin and ^{32}P-NAD$^+$ and they appeared to lack the 45-K molecular weight substrate. When this protein was purified from another source and then reconstituted into membranes of cyc$^-$ cells, they regained the sensitivity of their adenylate cyclase to hormones. This provided strong evidence that the 45-K protein represents a functional component of G$_s$. Attempts to purify G$_s$ from membrane preparations resulted initially in the identification of two proteins which usually co-purified together. One of these was the 45-K molecular-weight protein described above, while the second was a smaller molecule (35 K) which was also essential to the functional activity of G$_s$. These were designated as α- and β-subunits, respectively. Subsequently, a third protein component was also identified as part of the complete G$_s$ system, being a subunit of 8–10-K molecular weight which has now been designated γ. The structure and function of these components are described in more detail in Chapter 3.

Activation of adenylate cyclae

Kinetic studies of the activation of adenylate cyclase by hormones have suggested that the hormone receptor undertakes a catalytic role in the overall process. Thus, the binding of an agonist to a stimulatory receptor results in the formation of a complex which dissociates rather slowly relative to the time required for the receptor to encounter and activate G$_s$. Since this latter step is transient, the receptor has the capacity to interact with several different G$_s$ molecules during the time it contains a single molecule of bound hormone. This is therefore one of the steps at which amplification of the hormonal signal can occur during transduction.

The interaction of an occupied receptor with G_s promotes the dissociation of bound GDP and its replacement by GTP, thereby facilitating activation of G_s. Completion of the GDP–GTP exchange reaction causes dissociation of the G_s complex, such that $G_{s\alpha}$ is released with its bound GTP while the $\beta + \gamma$ subunits apparently remain associated together. $G_{s\alpha}$–GTP can then interact with the catalytic subunit of adenylate cyclase to promote activation and cAMP formation. The whole process is terminated by the intrinsic GTPase activity of $G_{s\alpha}$ which hydrolyses the γ-phosphate of GTP to yield GDP. In this form, cyclase activation is no longer favoured and $G_{s\alpha}$ re-associates with the free $\beta + \gamma$ subunits to await a further encounter with a hormone-occupied receptor. Thus, the cycle is completed. This model is illustrated in Fig. 4.3 and forms the basis of the 'collision-coupling' hypothesis of adenylate cyclase activation. The model envisages that the separate components of the system can all interact functionally without the need for the formation of a single complex comprising all of the individual molecular species. Hence, it has been possible to study the activation of G_s by β-adrenergic receptors in a reconstituted system containing no catalytic subunit. However, this view of the components of the system as entirely separate entities which do not associate together within the plasma membrane may be an over-simplification of the true situation in cells. Indeed, the biochemical evidence does not entirely favour this idea since G_s and the catalytic subunit (C) appear to exist in close association during purification procedures and the stoichiometry of the complex formed by these molecules is 1:1. Furthermore, this ratio remains unaltered irrespective of whether G_s contains bound GTP or GDP. Therefore, a model which appears to most closely satisfy both the kinetic and the biochemical evidence is one in which the rate-limiting step represents the association between a hormone-activated receptor and G_s containing bound GDP. This interaction then promotes GTP–GDP exchange, subunit dissociation and results in immediate activation of a closely-associated catalytic subunit.

It might be envisaged from this model that once $G_{s\alpha}$–GTP has been converted to $G_{s\alpha}$–GDP by the GTPase, all that would be required for a second round of activation is receptor-mediated guanine-nucleotide exchange. This also appears to be an over-simplification since the $\beta + \gamma$ subunits are absolutely required in order to facilitate GTP–GDP exchange. Therefore, once GTP hydrolysis has occurred, it is obligatory for complete reformation of the G_s holoprotein to occur before a second round of activation can commence.

Hormonal inhibition of adenylate cyclase

In addition to the many hormones which activate adenylate cyclase in cells, there is also a whole family of hormones and neurotransmitters which attenuate the enzyme activity. These include certain opioid peptides, adrenaline (acting at α_2-adrenergic receptors) and somatostatin. The effects of

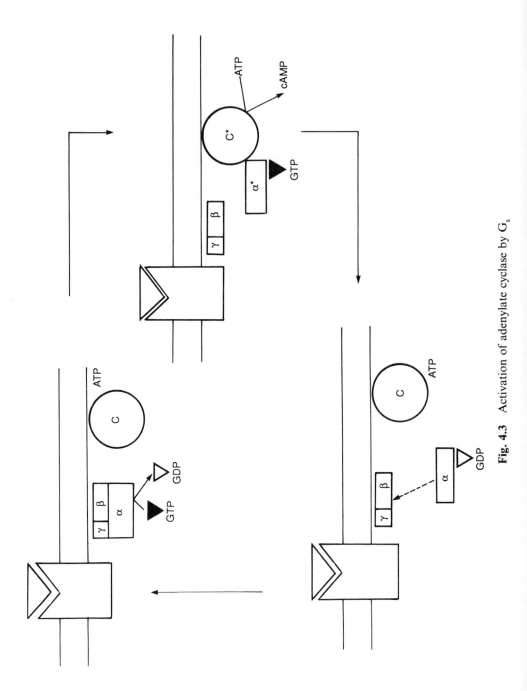

Fig. 4.3 Activation of adenylate cyclase by G_s

these agents are all mediated by a receptor-dependent mechanism, and it has become clear that their actions are also mediated by a guanine-nucleotide-binding protein. Suggestions that this might be the case first arose as a result of observations that the activity of adenylate cyclase in membrane preparations could be either enhanced or inhibited by GTP according to the concentration used. In general, the inhibitory effects required higher GTP concentrations than are needed for stimulation, but these data, coupled with the observation that it is possible to selectively modify either the stimulatory or the inhibitory component (with trypsin for example), led to the idea that a separate G-protein may be coupled to adenylate cyclase in an inhibitory fashion. This concept was strengthened by the realization that a toxin produced by the whooping cough bacterium, *Bordetella pertussis*, and originally characterized as an agent which prevented inhibition of insulin secretion by adrenaline, acts like cholera toxin to covalently modify a component of the adenylate cyclase system. As with cholera toxin, this modification was also found to be an ADP-ribosylation, but the substrate protein was of somewhat lower molecular weight than that modified in response to cholera toxin (42 K versus 45 K). Furthermore, exposure of cells to pertussis toxin could prevent the inhibition of adenylate cyclase by hormones. The pertussis toxin substrate was termed G_i. Like G_s, G_i has now been purified in homogeneous form and it is also composed of three subunits. The largest of these ($G_{i\alpha}$) is the target for ADP-ribosylation catalyzed by pertussis toxin and the remaining two subunits are very similar to the $\beta + \gamma$ subunits of G_s. In fact, peptide mapping and immunological studies suggest that $\beta + \gamma$ may be common to both proteins, although other indications suggest possible slight differences between the γ-subunits. 'Activation' of inhibitory receptors by appropriate hormones or drugs leads to dissociation of G_i into its constituent subunits and to stimulation of an intrinsic GTPase activity. The regulation of adenylate cyclase by both G_s and G_i, therefore represents a 'symmetrical' system having certain common components ($\beta + \gamma$) on each arm (Fig. 4.4). Interestingly, the available evidence

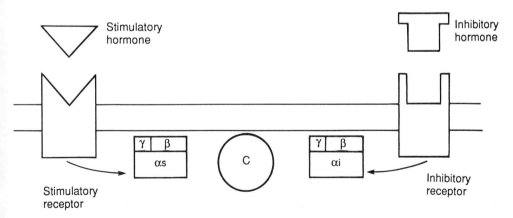

Fig. 4.4 Symmetry in the adenylate cyclase system

suggests that G_i is much more abundant than G_s in many cells, and this implies that G_i may potentially restrict the activity of adenylate cyclase even in the 'basal' state.

Mechanism of inhibition of adenylate cyclase by G_i

The molecular details of the mechanism(s) by which activation of G_i causes attenuation of adenylate cyclase activity have not yet been entirely clarified. Dissociation of the protein into its constituent subunits appears to be a necessary step and the exchange of GTP for bound GDP which precedes dissociation is stimulated by occupation of the receptor with agonist. ADP-ribosylation of G_i by pertussis toxin prevents hormone-induced dissociation of G_i, thereby affecting the hormonal signal. Studies performed with platelets have produced results which do not entirely support a direct inhibition of the catalytic subunit by $G_{i\alpha}$, and have led to the proposal of an indirect mechanism for inhibition of enzyme activity. This model relies on the finding that both G_s and G_i share two common subunits, $\beta + \gamma$, and that G_i is present in stoichiometric excess, relative to G_s in most cells. The proposal states that activation of G_i leads to subunit dissociation and the release of sufficient quantities of the $\beta + \gamma$ subunits to disturb the equilibrium position which exists between undissociated and dissociated G_s under resting conditions. It is proposed that a large increase in the amount of free $\beta + \gamma$ subunits in the membrane would (by mass action) drive the dissociation of G_s in the reverse direction to favour inactivation of the molecule. This model (Fig. 4.5), therefore, envisages that G_i will only be effective under conditions where the activity of adenylate cyclase is directly attributable to the level of dissociated $G_{s\alpha}$. In cells where this is not the case (for example in the mutant S49 cyc⁻ cells which lack functional $G_{s\alpha}$) the model predicts that receptors which are coupled to the enzyme in an inhibitory manner will not be functional. This theoretical prediction is not, however, borne out by experimental results, since inhibitory agonists are still capable of reducing adenylate cyclase activity in S49 cyc⁻ cells.

The most likely situation is that a combination of (at least) two mechanisms brings about hormone-mediated attenuation of adenylate cyclase. One of these involves direct inhibition of the catalytic subunit by $G_{i\alpha}$ (although direct evidence for this is still lacking) and the second is achieved by regulation of the dissociation state of G_s, via the release of $\beta + \gamma$ subunits from G_i.

Regulation of stimulatory receptors by G_i

Recently, evidence has begun to emerge which indicates that the role of G_i to limit activation of adenylate cyclase may also extend in another, previously unsuspected, direction. Surprisingly, recent results point to a direct effect of G_i upon stimulatory receptors themselves. The existence of this interaction thus requires us to consider yet another possible level of regulation in the already

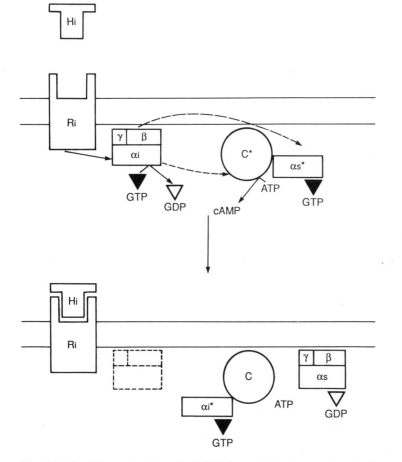

Fig. 4.5 Possible mechanisms for inhibition of adenylate cyclase by G_i

complex adenylate cyclase system. The observations have arisen out of experiments designed to examine whether purified hormone receptors can interact with different types of G-protein in a reconstituted membrane system. Surprisingly, it was found that the presence of purified avian β-receptors in the membrane can influence the extent to which G_i will bind guanine nucleotides. This suggests that G_i may be able to directly interact with stimulatory receptors, although the efficiency of this coupling is much less than the efficiency with which these receptors couple to G_s. Nevertheless, the possibility is raised that another mechanism by which hormones and neurotransmitters inhibit adenylate cyclase occurs as a result of direct attenuation of the extent of receptor–G_s coupling, mediated by G_i. Therefore, activation of G_i caused by the binding of an agonist to an inhibitory receptor, may lead to effects on the catalytic subunit of adenylate cyclase; the extent of dissociation of G_s and the efficiency of coupling of stimulatory receptors to G_s (Fig. 4.6).

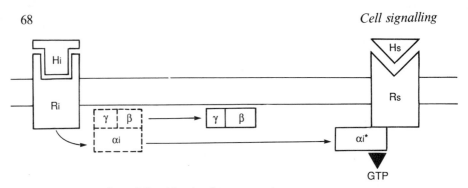

Fig. 4.6   Interaction of G_i with stimulatory receptors

It is clear, therefore, that the molecular mechanisms involved in the control of adenylate cyclase by hormones are considerably more complex than was originally envisaged by Sutherland. Regulation occurs at several stages in the overall sequence of events and the interrelationships between the various components of the system allow for rapid and sensitive responses at the cellular level.

The catalytic subunit of adenylate cyclase

The major enzymic component of the adenylate cyclase system is the catalytic subunit itself, which utilizes ATP in the formation of cAMP and pyrophosphate (Fig. 4.7). Isolation and purification of this molecule has proved to be an extremely difficult task, due largely to its relatively low abundance in cells and to its extreme lability when removed from its native membrane environment. In eukaryotic cells the enzyme exists exclusively in a membrane-bound form since, unlike guanylate cyclase, it is not found in the soluble fraction upon cell disruption. In an attempt to purify adenylate cyclase, use has recently been made of the selective interaction between the catalytic moiety of the enzyme and a diterpene compound isolated from Coleus plants, forskolin (Fig. 4.8). This agent has the capacity to directly activate the catalytic subunit without the intervention of either a receptor or a G-protein and has been used as an affinity ligand after coupling to Sepharose beads. Detergent extracts of heart membranes were applied to affinity columns containing the forskolin–Sep-

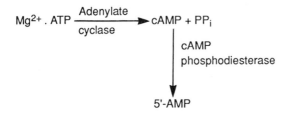

Fig. 4.7   Formation and metabolism of cAMP

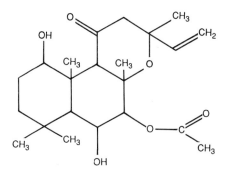

Fig. 4.8 The structure of the adenylate cyclase activator forskolin

harose resin and a fraction was eluted which could then be further fractionated by high-performance liquid chromatography to yield the purified enzyme. The final product was purified approximately 60 000-fold relative to the original extract and contained two major protein components. One of these could be labelled with cholera toxin and ^{32}P-NAD$^+$ suggesting that it was the α-subunit of G_s. This association between $G_{s\alpha}$ and C which was maintained throughout various rigorous purification steps, supports the idea that these two components must exist as a closely apposed complex in the cell. The second major protein was a larger molecule (molecular weight 150 K) which is believed to possess the catalytic activity. Further characterization of the protein will be required before it is possible to understand its mode of regulation in the cell, and its availability in pure form will now permit detailed reconstitution experiments to aid in this objective. One interesting, and rather unexpected, finding is that the protein binds to a lectin isolated from wheat germ which has affinity for specific carbohydrate residues. Hence, the catalytic subunit of adenylate cyclase may be a glycoprotein. Since glycoproteins are normally restricted to the external face of the plasma membrane, this raises the interesting possibility that adenylate cyclase could be a transmembrane enzyme. The functional significance of this orientation has not been determined, but the active site of the enzyme must be located on the cytoplasmic surface of the membrane since this is where cAMP formation occurs.

Regulation of adenylate cyclase by cations

Adenylate cyclase is an enzyme which requires metal ions for activity and in the intact cell Mg^{2+} is likely to fulfil this requirement. Part of the dependence of the enzyme on Mg^{2+} ions probably reflects the fact that the substrate ATP must be present as a Mg.ATP complex in order to allow cAMP formation. However, the discovery that Mg^{2+} ions, when added in excess of the ATP concentration, can induce a further increase in the V_{max} of the reaction suggests that Mg^{2+} may also play an allosteric role in enzyme activation. In *in vitro*

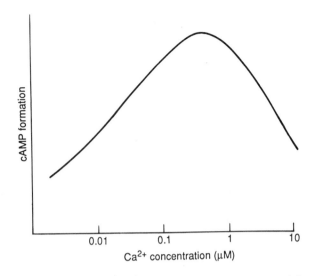

Fig. 4.9 Effect of Ca^{2+} ions on adenylate cyclase activity

assay systems, Mn^{2+} ions can also support adenylate cyclase activity but this is not likely to be of any physiological significance since the endogenous Mn^{2+} concentration in cells is very low. Ca^{2+} ions cannot directly substitute for Mg^{2+} as mediators of cyclase activity but, despite this, it has become evident that the enzyme from many tissues is sensitive to the prevailing Ca^{2+} concentration. The dose–response relationship which exists between the Ca^{2+} concentration and enzyme activity is complex, with low concentrations (submicromolar) of Ca^{2+} causing enzyme activation and higher concentrations (1–$10\,\mu M$) causing inhibition. The latter response may reflect competition between Ca^{2+} and Mg^{2+} for certain critical binding sites on the enzyme, and could play a role in the control of adenylate cyclase activity in the cell where free Ca^{2+} concentrations in this range can occur following cell stimulation.

Calmodulin as a regulator of adenylate cyclase

Calmodulin is a small ubiquitous protein which is responsible for mediating the effects of Ca^{2+} on many enzymes and has been implicated as a regulator of adenylate cyclase. The protein has the capacity to bind Ca^{2+} as the concentration of the ion increases, and this results in a change in conformation which leads, in turn, to enzyme activation. A large proportion of the calmodulin present in cells is associated with membranes (up to 50%), but if membrane preparations are treated with chelating agents (such as EGTA) to remove endogenous Ca^{2+}, the calmodulin content can also be depleted. When this is done, the basal activity of adenylate cyclase, the extent of stimulation mediated by hormones and the activation by Ca^{2+} are all reduced. If, however,

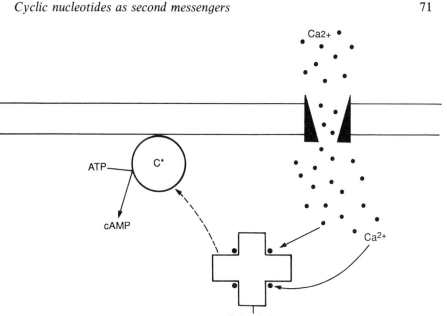

Fig. 4.10 Activation of adenylate cyclase by Ca^{2+}-calmodulin

exogenous calmodulin is then added back to these preparations, these responses can be recovered. Thus, it appears that calmodulin may play a central role in the regulation of adenylate cyclase activiy in the cell (Fig. 4.10).

The molecular basis of the control of adenylate cyclase by calmodulin has not been unequivocally determined. In particular, it remains unclear whether the effect of the modulator reflects a direct interaction with the catalytic subunit of the enzyme, or whether control is mediated through the G-protein system. Evidence has been obtained in brain tissue that calmodulin can exert direct effects on the catalytic subunit, but other studies have also revealed that complex interactions exist between calmodulin and guanine-nucleotide analogues in activation of adenylate cyclase. It is possible, therefore, that calmodulin may interact with several components of the adenylate cyclase in order to exert its overall effect.

It is attractive to postulate that stimulation of cells by agents which act primarily to raise the cytosolic-free Ca^{2+} concentration would, in consequence, induce a rise in cAMP by activation of adenylate cyclase by the mechanism outlined above. Indeed, this does appear to happen in some tissues. For example, in islets of Langerhans the major physiological stimulus, glucose, causes a modest Ca^{2+}-dependent rise in intracellular cAMP which has been attributed to the calcium-calmodulin-dependent activation of adenylate cyclase. However, such responses do not always directly correlate with the measured Ca^{2+} dependency of adenylate cyclase in tissue extracts. In general, the Ca^{2+}-mediated activation process occurs at concentrations of Ca^{2+} which are at, or below, the resting level for Ca^{2+} found in cells ($\sim 0.1\,\mu M$). Raising the Ca^{2+} concentration into the range $1-10\,\mu M$, which is equivalent to the rise induced by agonists in many cells, actually leads to inhibition of adenylate cyclase activity. This seems surprising but may be the more physiologically relevant effect since at these higher Ca^{2+} concentrations, calmodulin-dependent activation of cAMP phosphodiesterase also occurs which will tend to increase the rate of cAMP degradation. It would seem inefficient for a cell to have a system which simply increases the rate of cAMP turnover at elevated intracellular Ca^{2+} concentrations and thus, this concerted 'switch-off' mechanism which operates to decrease cAMP as Ca^{2+} rises, may serve to limit the extent of any possible synergistic interactions between the two messengers.

Guanylate cyclase and cGMP

Guanylate cyclase was first identified in animal tissues in the late 1960s and it soon became apparent that, unlike adenylate cylase, it exists in two distinct forms, one of which is membrane-bound and the other soluble (Fig. 4.11). In tissues such as platelets the activity is located almost exclusively in the cytosol whereas in kidney it is predominantly particulate. In most tissues, both forms of the enzyme are present, although to varying extents. The two forms of the enzyme have different properties (both physicochemical and kinetic) and can be distinguished immunologically, suggesting that they are completely different molecules. The soluble enzyme has been the better studied of the two forms due to its relative ease of isolation and purification. Conversely, it is the membrane-bound enzyme that is most likely to act as a transducer of extracellular signals in cells.

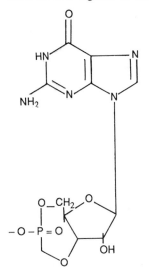

Fig. 4.11 The structure of cGMP

Soluble guanylate cyclase

The soluble isozyme of guanylate cyclase is a large protein (M_r 150 K) which is composed of two non-identical subunits and requires divalent cations for activity. This role is probably fulfilled by Ca^{2+} and Mg^{2+} in intact cells, although Mn^{2+} will support enzyme activity *in vitro*. The enzyme has a K_m for GTP of $\sim 30\,\mu M$ but requires quite high levels of cations, having K_m for Mg^{2+} in the millimolar range. Much effort has been expended to demonstrate that extracellular ligands can modulate soluble guanylate cylase activity but this has proved a rather fruitless exercise so far. Indeed, it is perhaps not surprising that the soluble form of the enzyme (which is not directly accessible to hormone receptors) is not a direct target for hormones, although it has been reported that the enzyme is activated by arachidonic acid in some cells which may provide a link between hormones acting at the cell surface and intracellular signalling systems. A number of other agents can also alter the enzyme activity under appropriate conditions (e.g. the oxidized form of vitamin C, dehydroascorbic acid, and several nitrosamines can activate the enzyme) which may reflect free-radical generation, since guanylate cyclase can be activated by agents that form free radicals. Indeed, this may be the mechanism by which sodium nitroprusside, a commonly used activator of the enzyme, brings about its effects.

Particulate guanylate cyclase

The membrane-bound form of guanylate cyclase has received a considerable revival of attention recently since it has been found that a special group of hormone-like peptides (atriopeptides) may bring about their effects by activation of this enzyme. For this reason, these molecules represent a unique group of hormones, as no other endogenous ligands which work by this mechanism have so far been identified. It is noteworthy, however, that a toxin produced by pathogenic strains of *Escherichia coli* may activate the enzyme in gut mucosa resulting in ionic imbalance and water loss. The atriopeptides are a group of molecules with a common function which relates to the promotion of smooth-muscle relaxation, diuresis and excessive sodium excretion from the kidney. The most active species is a 28-amino acid peptide which is produced predominantly in the right atrium of the heart, and is active on vascular smooth muscle and on the kidney. The observed ability of the peptide to induce sodium excretion (natriuresis) together with its site of synthesis in the heart, have led to it being named atrial natriuretic peptide (ANP). A crucial observation that has provoked particular attention to the mechanism of signal transduction of this peptide was that treatment of rats with ANP results in a marked increase in the concentration of cGMP measured in the blood and urine. This led to the idea that ANP may control the activity of guanylate cyclase in its target cells.

Receptors for ANP have been identified by ligand-binding experiments using radiolabelled derivatives of the peptide as probes. The major surprise

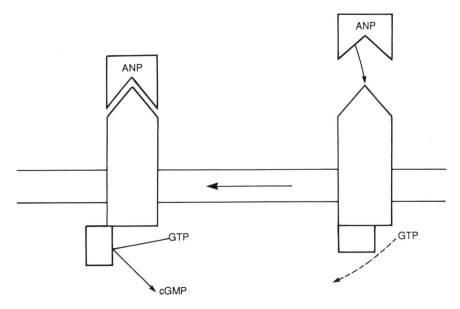

Fig. 4.12 Possible mode of activation of particulate guanylate cyclase by atrial natriuretic peptide

came when attempts were made to purify these receptors, since it became clear that they are very tightly associated with a particulate form of guanylate cyclase. Indeed, isolation of the receptor from rat lung and from adrenal cortex resulted in the preparation of a protein which could not be distinguished from the receptor itself. In the case of the lung, the purified receptor was a large glycoprotein of $M_r \sim 120\,K$, whereas in adrenal cortex it was a protein of $M_r \sim 180\,K$. However, in both cases the receptor co-purified with guanylate cyclase, which points to the remarkable possibility that the guanylate cyclase catalytic activity may reside in the *same protein* as the receptor binding unit (Fig. 4.12). Recently, the gene coding for a membrane form of guanylate cyclase has been isolated from rat brain. This gene was cloned and transfected into cultured cells which were then shown to possess an active guanylate cyclase. Significantly, these cells also acquired the ability to bind ANP in a manner that resulted in enhancement of guanylate cyclase activity. These results demonstrate conclusively that the receptor and the catalytic activity reside in the same protein. The gene codes for a protein which is very large (1057 amino acids; M_r 116 K) and contains a single transmembrane domain. This divides the sequence such that approximately 45% of the amino acids are in the extracellular space (where the binding site for ANP is located) while about 55% are intracellular (and contain the catalytic site).

If the scheme described above proves to be a correct interpretation of the data then this automatically implies that a new mechanism of signal transduction has been discovered. It has been known for some time that particulate guanylate cyclase is a transmembrane protein and these new results suggest that it may contain a hormone binding site within its structure.

Hence, enzyme activation could be achieved by an allosteric mechanism which directly transmits a change in conformation through the membrane to the catalytic site. If this is the case, then guanylate cyclase may be controlled in a manner similar to that initially envisaged by Sutherland for adenylate cyclase, and which proved to be an over-simplification!

Receptor-mediated activation of guanylate cyclase has also been observed in the spermatozoa of sea urchins, where the active peptides are derived from the eggs of these animals. In such invertebrates the receptor protein for the peptides also co-purifies with guanylate cyclase in certain species, suggesting a direct relationship between the two. In this situation, peptide binding is associated with enzyme activation, although the response is transient. It has been found that the secondary loss of enzyme activity coincides with a decrease in the extent of phosphorylation of the enzyme suggesting that it may be regulated by a phosphorylation/dephosphorylation mechanism. The proposal has been made that the guanylate cyclase may exist in a phosphorylated form under basal conditions, and that this allows for efficient activation upon ligand binding. The conformational change induced by ligand binding then exposes the phosphorylation site which is rapidly dephosphorylated (by a phosphatase) and leads to a reduction in enzyme activity. This would, therefore, represent a 'switch-off' mechanism operating after receptor activation.

This scheme is relatively speculative at present, but if similar mechanisms are found to be present in mammals, then the guanylate cyclase system may yet reveal a regulatory complexity which at least matches that of adenylate cyclase!

Mechanism of action of cGMP

Until recently cGMP was a very enigmatic molecule – indeed, to some extent it remains so – although a clearer picture of the role played by cGMP as an intracellular messenger is now emerging. When cGMP was discovered in the early 1960s, the role of cAMP as an intracellular second messenger was already becoming widely appreciated. It is not surprising, therefore, that the discovery of a very similar molecule in cGMP was greeted with enthusiastic optimism that the elucidation of a new messenger system would soon follow. In retrospect, this has proved to be an over-optimistic belief since, despite considerable effort, little convincing evidence has been forthcoming to support a generalized second messenger role for cGMP. However, recent studies have shown that it may fulfil specialized functions in certain tissues that at least one type of extracellular signal molecule, atrial natriuretic peptide, uses the cGMP-system to transduce its receptor interaction into a physiological response (see above).

The role of cGMP in visual transduction

Surprisingly, one tissue in which cGMP has emerged as a central regulatory molecule is the photosensitive rod cell of the vertebrate retina. In this tissue, the soluble form of guanylate cyclase is very active and the concentration of cGMP is high under resting conditions (about 50 μM). This seems anomalous if the cyclic nucleotide is to function as a second messenger since it is usual to think in terms of an increase in messenger concentration as a necessary prelude to the activation of a physiological response. However, the characteristics of the rod cells are unusual in that they respond to a *decrease* in cGMP rather than to a rise in its concentration. This has now been explained by the important discovery that the rod-cell plasma membrane contains an ion channel which is maintained in the open state under non-stimulating conditions (i.e. in the dark). This channel seems to be relatively non-specific in that it allows the inward passage of both Na^+ and Ca^{2+} and ensures that the plasma membrane is maintained in a depolarized state when the cell is not illuminated. In order to elicit the transmission of a nerve impulse, the rod cell has to become hyperpolarized, an effect which requires the closure of the ion channels. The crucial observation which has helped to elucidate the bio-chemistry of this process is that cGMP controls the opening of this channel. This has been demonstrated both in intact cells where micro-injection of cGMP causes a 20-fold enhancement of the inward current in dark-adapted cells, and in using the patch-clamp technique in which the current flow through small regions ('patches') of excised membrane can be stimulated by appli-cation of cGMP. Analysis of the electrophysiological properties of the channel has revealed that cGMP increases the probability of channel opening and that this effect depends upon the cube of the cGMP concentration. This has been taken as evidence that the channel possesses three binding sites for cGMP, all of which must be occupied in order to facilitate channel opening. The important point about this mechanism is that it is a direct effect of cGMP itself; phosphorylation reactions do not play any role in the response. The binding site on the channel shows marked specificity for cGMP, since both cAMP and cIMP are inactive as channel openers. Furthermore, a number of modified analogues of cGMP are also inactive (e.g. dibutyrl-cGMP, deoxy-cGMP) whereas others (e.g. 8-bromo-cGMP) are more potent than cGMP itself and it has been inferred from such data that hydrogen bonding is an essential component of the binding reaction.

The sequence of events leading to activation of the rod cell is as follows (Fig. 4.13):

1. A photon of light is absorbed by rhodopsin which then promotes dissociation of the G-protein transducin into its constituent subunits.
2. The active α-subunit of transducin interacts with, and activates a specific cGMP phosphodiesterase.
3. cGMP is degraded and its concentration in the cell falls, leading to closure of the plasma membrane ion channels.
4. The cell becomes hyperpolarized due to reduced Na^+ and Ca^{2+} influx which promotes synaptic neurotransmitter release and nerve impulse propagation.

From this scheme it is evident that the role of the cGMP phosphodiesterase

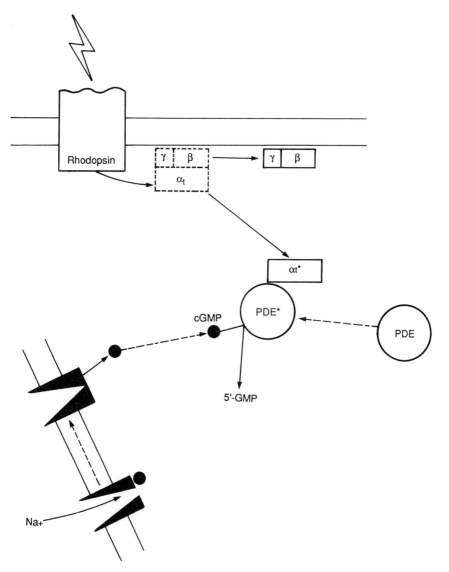

Fig. 4.13 Control of plasma membrane ion flow by cGMP in the rod cells of the retina

is to activate signal transmission rather than to terminate it, as is usually the case in cells which use cAMP as an intracellular messenger.

cGMP phosphodiesterase

The rod-cell cGMP phosphodiesterase has been extensively characterized, not least because it represents an intracellular enzyme whose activity is regulated by a G-protein, transducin. The enzyme from many vertebrate species is a large (180-K) molecule composed of three types of subunit (α, β, γ) in a probable stoichiometry of 1:1:1. The γ-subunit is the smallest component (~ 11 K) and is an inhibitory subunit which regulates the catalytic activity of the $\alpha + \beta$ subunits. Inhibition can be as great as 99% and removal of the γ-subunit leads to as much as 80-fold activation of the phosphodiesterase activity. The α- and β-subunits of the molecule are closely related in structure and are both approximately 85 K in molecular weight. They are capable of hydrolysing cGMP at a very rapid rate, having a combined turnover number as high as $3500\,\text{s}^{-1}$ and a K_m for cGMP of $\sim 0.1\,\text{nM}$. Indeed, it has been calculated that the absorption of a single photon of light can lead to the hydrolysis of 100 000 molecules of cGMP under optimal conditions!

It is tempting to speculate that transducin promotes activation of the phosphodiesterase by causing removal of the γ-subunit from the holoenzyme. There is evidence to support this view but the precise molecular details which underlie the process remain obscure.

The photosensitive rod cell, therefore, exhibits a novel cyclic nucleotide-mediated signal transduction system which relies on direct effects of cGMP and does not involve protein kinase activation. The extent to which this mechanism has been diversified for use in other electrically-active (and perhaps non-excitable) tissues is now of great interest and evidence is already being gathered which suggests that it may be of more widespread significance. Indeed, some results point to the possible existence of cAMP-regulated channels in the cells responsible for detection of olfactory stimuli.

Cyclic nucleotide phosphodiesterase

In order to maintain the sensitivity of a second messenger system it is necessary to have a mechanism which allows any rapid increases in the messenger concentration to be compensated by an equally rapid decrease when the hormone is removed. Failure to do this would obviously result in a prolonged response to the hormone which would not then be suitable for mediating acute cellular changes. In 1958, Sutherland and Rall discovered an enzyme which was capable of hydrolysing cAMP to yield 5'-AMP, and was, thus, a 3':5' cyclic nucleotide phosphodiesterase. A variety of different forms of this enzyme have since been discovered, some of which are present in the cytosol while others are associated with cellular membranes. It is now clear that these enzymes represent the only route by which cAMP and cGMP can be degraded in cells.

Early studies revealed the presence of three distinct forms of phosphodiesterase in several tissues, which differed with respect to their molecular size and properties. Some of these had a higher affinity for cAMP than for cGMP, while others showed the reverse specificity. Many studies employed ion-exchange chromatography to fractionate the various forms of the enzyme and the three major types were designated I, II and III to denote their order of elution from DEAE-cellulose columns. The rates of hydrolysis of cAMP and cGMP vary between these preparations and types I and II have a lower K_m for cGMP than for cAMP. In many tissues, these two forms of the enzyme are also Ca^{2+} dependent, although the extent of activation by Ca^{2+} is variable.

Electrophoretic methods have also been used to fractionate phosphodiesterase preparations and as many as seven individual forms have been identified using this method. It is difficult to definitively equate particular fractions obtained by one method with those found by another, but it is clear that most cells contain multiple forms of phosphodiesterase and that these are differentially compartmentalized and have variable substrate specificities.

In general, it appears that hormones and receptor agonists do not regulate the concentration of cAMP by directly controlling the activity of phosphodiesterase enzymes; hence, the major determinant of the cAMP concentration in a cell is the adenylate cyclase activity. However, there are some notable exceptions to this generalization, including regulation of the cGMP concentration by light in the rod cells of the retina (see above) and control of hepatic cAMP levels by insulin. In both of these cases, an important component of the signalling mechanism may involve phosphodiesterase activation. One reason why this mechanism is not more widespread may be that the rate of flux through the cAMP pathway is quite low under resting conditions and, therefore, changes in phosphodiesterase activity will have relatively little effect. In this context it is noteworthy that a number of pharmacological agents which inhibit phosphodiesterase activity have been described, but these compounds only have striking effects on cell cAMP levels under conditions when adenylate cyclase is simultaneously activated. Furthermore, the physiological mechanisms which have been described all result in phosphodiesterase activation and thus operate to lower cyclic nucleotide levels. No cellular mechanism has yet been described whereby phosphodiesterase activity is inhibited. Hence, control of phosphodiesterase activity always occurs in situations where a lowering of the second messenger concentration is required.

Perhaps the best studied of the phosphodiesterase enzymes are those whose activity is regulated by Ca^{2+}, and these will serve as suitable examples to illustrate the function of these molecules.

Ca^{2+}-dependent cyclic nucleotide phosphodiesterase

Many tissues contain Ca^{2+}-dependent phosphodiesterases which can be classified as type I or type II enzymes according to their chromatographic properties on ion-exchange resins. These enzymes occur in the soluble fraction of the cell and have widely differing K_m values for cyclic nucleotides depending on their tissue of origin. In general, they have a higher affinity for cGMP ($K_m \sim 0.1\text{–}200\mu M$) than for cAMP ($K_m \sim 5\text{–}500\mu M$), although the V_{max} is greater with cAMP as substrate. They require Mg^{2+} ions for activity and their sensitivity to Ca^{2+} depends upon interaction with the regulatory protein calmodulin. Indeed, many assay systems for calmodulin are based on its ability to activate cyclic nucleotide phosphodiesterase in the presence of Ca^{2+}.

Under basal conditions, calmodulin is believed to be present in cells as a

Mg^{2+}-calmodulin complex but as the concentration of Ca^{2+} rises into the micromolar range, Ca^{2+} displaces Mg^{2+} from the binding sites to yield a $Ca^{2+}_{(3)}$–$Mg^{2+}_{(1)}$–calmodulin complex. In this form, calmodulin then binds to and activates phosphodiesterase. The precise stoichiometry of this activation is still in doubt but the reaction is immediately and completely reversible when the Ca^{2+} concentration falls. The increase in enzyme activity mediated by Ca^{2+}–calmodulin has been attributed to a decrease in the K_m for cyclic nucleotides, although changes in V_{max} have also been noted in the enzyme from some tissues.

It is evident from the foregoing that although most hormones do not control the activity of phosphodiesterase directly, they can do so by the indirect mechanism of regulating cellular Ca^{2+} handling. Since a number of extracellular signals do indeed alter the concentration of cytosolic Ca^{2+} in cells, it follows that these can also regulate the cAMP level. Since calmodulin is present in cells in a large excess relative to phosphodiesterase it is likely that fluctuations in the Ca^{2+} concentration are the primary determinants of the enzyme activity. As discussed previously, some cells also contain Ca^{2+}–calmodulin-sensitive adenylate cyclase which raises the possibility that a rise in cytosolic Ca^{2+} may promote an increase in both the rate of synthesis and degradation of cAMP under certain circumstances. Suggestions have been made that these two actions could occur sequentially, as a consequence of the spread of a wave of influent Ca^{2+} through the cell from its site of entry at the plasma membrane. However, the different Ca^{2+} sensitivities of the two enzymes make it likely that the phosphodiesterase is the primary target for regulation by Ca^{2+}.

Insulin-stimulated phosphodiesterases

Insulin is a hormone which attenuates cAMP levels in some cells (e.g. hepatocytes, adipocytes) a response which may reflect direct inhibition of adenylate cyclase, but is also related to activation of particular phosphodiesterase enzymes. Using a rapid fractionation method, it has been possible to identify two distinct forms of phosphodiesterase in liver which are activated by insulin. One of these is an enzyme which is associated with the plasma membrane, while the second appears to reside in an intracellular membrane-bound compartment and has been termed the 'dense-vesicle' enzyme. These two forms of phosphodiesterase differ in terms of their physical and immunological properties and they appear to be unrelated species. There is some evidence that activation of the two enzymes by insulin may also occur by different mechanisms although the molecular details of these processes remain obscure. Furthermore, the respective physiological roles of the enzymes also remains in doubt. It has been reported that insulin can still effectively lower glucagon-stimulated cAMP levels under conditions where activation of the 'dense-vesicle' enzyme is prevented, which seems to suggest that the plasma-membrane enzyme is the more important for control of hepatic cAMP levels by insulin. It remains to be established how insulin controls the activity of this enzyme; Ca^{2+} is not thought to be involved but there are suggestions that changes in the phosphorylation state of the protein may be important.

The effector system for cAMP

In order to exert its effects in the cell, cAMP must interact with, and activate, an effector system which will convert the hormone signal into a cellular response. In the 1960s several research groups were attempting to locate this effector system and succeeded in identifying a protein which could act as a cAMP receptor. This protein bound cAMP had high affinity and sufficient specificity to suggest that it could function as an intracellular receptor under physiological conditions. The molecule was later identified as part of an enzyme which had previously been shown to act as a protein kinase whose activity was stimulated by cAMP. It soon became apparent that this molecule occurs ubiquitously in mammalian cells and it was proposed as the mediator of all cellular actions of cAMP. Indeed, this proposition remains valid and it is now generally held that all actions of cAMP occur at the intervention of this enzyme. The properties of cAMP-dependent protein kinase (PK-A) itself have been extensively studied and are considered in more detail in Chapter 8.

Evidence that PK-A does, indeed, represent the cellular cAMP receptor was provided by studies which showed that the actions of cAMP could be mimicked by micro-injection of the purified protein kinase into cells. Purification studies also revealed that the enzyme is comprised of two types of subunit, one of which contains the binding sites for cAMP, while the second possesses the catalytic activity. In the cell, under basal conditions, these two types of subunit are tightly associated together in a tetrameric complex which contains two binding or regulatory (R) subunits and two catalytic (C) subunits (Fig. 4.14). Binding of cAMP to the R subunits leads to dissociation of the complex and release of the active catalytic subunits which are then available to phosphorylate substrate proteins.

Fractionation of PK-A on anion-exchange resins has revealed the presence of two forms of the enzyme, designated types I and II. These have identical catalytic subunits but immunologically-distinct R subunits. In some cells (e.g. heart) a large proportion of the type-II enzyme is associated with membranes although, in general, the enzymes are found in the cytosolic fraction of the cell. The presence of two forms of the enzyme has raised speculation that they may provide additional specificity to the cAMP system by being preferentially located in different cellular compartments. Furthermore, the ratio of the two types of enzyme appears to vary according to the phase of the cell cycle and has led to speculation that the type-I enzyme may be involved in the stimulation of cell growth. The finding that some tumour cells have an increased proportion of this form of the enzyme supports such an idea. One obvious difficulty with this concept, however, derives from the fact that the differences between types I and II of the PK-A lie in the regulatory subunits and not in those with catalytic

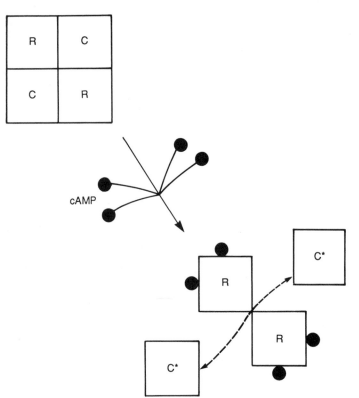

Fig. 4.14   Activation of protein kinase A by cAMP. R = regulatory subunit; C = catalytic subunit

activity. This could imply that their differential cellular compartmentation is the sole determinant of the range of protein substrates which are available for phosphorylation or that the R subunits have a separate functional role which is unrelated to the kinase activity. This remains an open question, but it is intriguing that the R subunits of the enzyme have been found in the nuclei of stimulated cells and that their translocation to this site appears to depend on the binding of cAMP. Furthermore, it has been established that cAMP can initiate an increase in the rate of transcription of specific genes (e.g. prolactin in pituitary cells and tyrosine aminotransferase in liver). The mechanisms by which this occurs have not been established and could still be related to changes in the phosphorylation state of particular chromatin proteins rather than to any direct action of the regulatory subunits themselves. It should soon be possible to answer this question since a cAMP-regulated gene has now been isolated from rat liver (the gene coding for phosphoenolpyruvate carboxy-kinase) and a sequence of DNA identified which confers sensitivity to the cyclic nucleotide. Part of this sequence bears similarity to that found in the genes of some bacteria which are also regulated by cAMP and indicates that it could have been conserved during the evolution of higher organisms.

Another potential role for the regulatory subunit of cAMP-dependent protein kinase type II has recently been suggested by the demonstration that, when purified, this molecule can inhibit a phosphoprotein phosphatase enzyme. This effect requires the presence of cAMP, suggesting that it results from the formation of the R–cAMP complex, and it has been observed both in cell extracts and with purified components. This effect would be of obvious physiological significance if it also occurs in the cell, since it would tend to amplify the effect of cAMP by reducing the rate at which phosphate groups are removed from substrate proteins during hormone stimulation.

Despite these findings with the type-II enzyme there is evidence that the type-I protein kinase is preferentially activated in cells which contain low concentrations of cAMP. In particular, chromatography of tissue extracts on a hydrophobic system which separates the two types of protein kinase has revealed that only the type-I enzyme exists in a dissociated state under basal conditions in liver cells. Furthermore, this enzyme becomes activated at lower cAMP concentrations than those that are required to cause dissociation of the type-II form. In addition, data have been presented which suggest that the two forms of the enzyme may be differentially activated by hormones in cells from a human breast tumour. The recent availability of cAMP analogues which selectively interact with each form of the enzyme should be of further value in unravelling the relative importance of each enzyme in cell physiology.

cGMP-binding proteins

A cGMP-dependent protein kinase has been identified in cell extracts from many mammalian tissues and appears to be a largely soluble protein which is composed of two subunits. Each of these can bind cGMP and both also seem to have catalytic activity. These subunits may be related to the cAMP-dependent protein kinase since they share certain similarities of substrate specificity and molecular shape and they also partially cross-react with antibodies to the cAMP-dependent enzyme. Rather less emphasis has been placed on the cGMP-dependent protein kinase than on that activated by cAMP, but it is clear that this enzyme can phosphorylate certain proteins in cells. The functional significance of these effects has still not been fully established.

Physiological regulation of protein kinase activity

In its non-dissociated, tetrameric, form the cAMP-dependent protein kinase does not have catalytic activity. This probably reflects the sites of interactions of the R and C subunits, which combine in such a manner that the active site on the C subunits is buried within the holoenzyme complex. Studies using cell extracts have revealed that, under basal conditions, approximately 75% of the

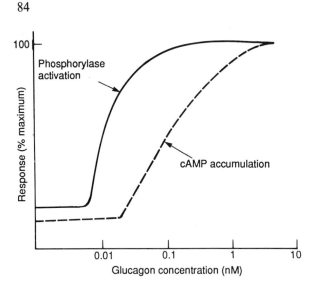

Fig. 4.15 Effect of glucagon on phosphorylase activation and cAMP accumulation in hepatocytes

protein kinase is present as the inactive tetramer and that this figure rapidly decreases as the concentration of cAMP rises. Quite small increases in cAMP are sufficient to induce a significant rise in protein kinase activity which allows marked amplification of the response to a hormone. Indeed, it has often been observed that a maximal physiological response can be elicited by a dose of agonist which produces much less than the maximal rise in cAMP. For example, in liver cells stimulated with glucagon, the maximum increase in phosphorylase activity occurs at a hormone concentration of 0.2–0.5 nM, whereas at least 20 times more hormone must be added in order to produce a maximal rise in cAMP concentration (Fig. 4.15). In most cells, a full physiological response can be induced by as little as a two-fold increase in cAMP, which correlates with an increase in the extent of dissociation of the protein kinase to approximately 50% of the total.

It is, perhaps, surprising that as much as 25% of the cAMP-dependent protein kinase is present in its dissociated form under resting conditions, but this may reflect the presence of a specific protein which inhibits the enzymes. Such a protein has been identified and purified and is sufficiently abundant in skeletal muscle and heart to completely inhibit the protein kinase under 'basal' conditions. This molecule is a heat-stable protein which interacts with the free catalytic subunits of the enzyme but does not bind to the holoprotein. The precise role in the control of cAMP-dependent processes has still to be clarified, as it is unclear whether the inhibitory activity of the protein is also subject to regulation.

Hormonal regulation of glycogen metabolism

One of the best studied systems in which cAMP acts as second messenger is the regulation of glycogen metabolism, and the essential features of this response are similar in both muscle and liver. It is well known that adrenaline promotes glycogen degradation in these tissues, and, in skeletal muscle, this contributes to the increased rate of glycolysis necessary for maintenance of ATP levels during contraction. Glucose release is achieved via a phosphorylation cascade which serves to amplify the hormone signal and is initiated by a rise in cAMP. The scheme is illustrated in Fig. 4.16.

Adrenaline binds to β-adrenoceptors on the surface of muscle cells and causes a rapid activation of adenylate cyclase. This leads, in turn, to a rise in intracellular cAMP (which can be detected within seconds of hormone addition) and activation of protein kinase A. One of the major substrates for this enzyme in skeletal muscle is a second protein kinase (phosphorylase b kinase) which shows specificity for the enzyme responsible for glycogen degradation. Phosphorylase kinase is a multi-subunit enzyme composed of four different types of subunit ($\alpha_4\beta_4\gamma_4\delta_4$), two of which ($\alpha$ and β) are phosphorylated on specific serine residues by the cAMP-dependent protein kinase. Phosphorylation of the β-subunits occurs most rapidly in the cell and is associated with activation of the enzyme; the α-subunit phosphorylation is slower and does not change the enzyme activity. Muscle phosphorylase exists in a dimeric form and each of the constituent subunits incorporates 1 mole of phosphate under the influence of phosphorylase kinase. This phosphate becomes covalently linked to a serine residue at position 14 of the amino acid sequence and results in stimulation of the enzyme activity. This is followed by the rapid release of glucose residues from the glycogen particles with which phosphorylase is associated in muscle.

In concert with the activation of glycogen breakdown, cAMP also inhibits the rate of glycogen synthesis by causing inactivation of glycogen synthase. This enzyme also exists in two forms which are interconvertible by phosphorylation. However, in contrast to the situation with phosphorylase kinase, the incorporation of P_i into the two identical subunits of this enzyme under the influence of PK-A results in a loss of catalytic activity. The net result is therefore, to increase the rate of glycogen breakdown. Glycogen synthase can be phosphorylated at several different sites by PK-A and can also be phosphorylated by phosphorylase kinase. This latter reaction also leads to partial inactivation of the enzyme but, interestingly, the phosphorylated residue (serine-7) is not one that serves as a substrate for the cAMP-dependent enzyme.

A third point at which cAMP controls muscle glycogen breakdown lies at the level of the enzyme which dephosphorylates phosphorylase a. This enzyme, phosphorylase phosphatase, has a molecular weight of 35 K and can dephosphorylate several substrates including phosphorylase, phosphorylase kinase and glycogen synthase. As such, it can counteract the effect of a rise in cAMP by 'switching-off' the enzymes involved in glycogen degradation.

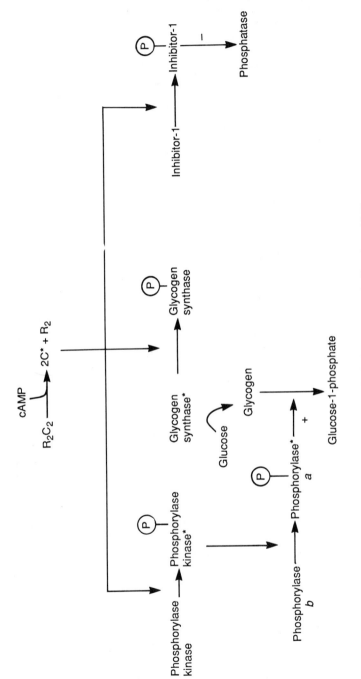

Fig. 4.16 Control of glycogen metabolism by cAMP

However, phosphorylase phosphatase is itself subject to regulation by yet another protein, inhibitor-1, which can cause inhibition of the phosphatase activity. Inhibitor-1 serves as a substrate for cAMP-dependent protein kinase and its ability to inhibit phosphorylase phosphatase is only apparent after the protein has been phosphorylated by this enzyme. Therefore, activation of protein kinase A leads to phosphorylation of inhibitor-1 and thereby causes inhibition of phosphatase activity. This then tends to amplify and prolong the activation of glycogen breakdown by maintaining all of the enzymes in their phosphorylated states.

This example demonstrates how a receptor agonist acting at the cell surface can, by raising the concentration of cAMP, coordinately regulate a complex sequence of reactions in the cell to produce a physiological response. This system also shows how effectively a second messenger can amplify the hormone signal. It has been estimated that each molecule of hormone can elicit the mobilization of some 1000 glucose residues from glycogen by utilizing this cascade mechanism.

cAMP and Ca^{2+} as second messengers

In the above example of a physiological process which can be regulated by cAMP, it is also clear that Ca^{2+} ions can serve a similar second messenger function. One of the four types of subunits present in phosphorylase kinase has been found to be identical with calmodulin and this confers on the enzyme a marked sensitivity to Ca^{2+}. Thus, a rise in Ca^{2+} concentration can completely activate the enzyme. Therefore, in muscle and liver, cAMP and Ca^{2+} have very similar actions and both control the breakdown of glycogen by inducing activation of phosphorylase kinase. However, in other cells the relationship between these two second messengers is less straightforward. For example, stimulation of insulin secretion from islets of Langerhans results from a rise in the concentration of Ca^{2+} in the islet B-cells. These cells also have receptors for cAMP-dependent hormones (eg. glucagon, adrenaline) and cAMP can cause an increase in the extent of phosphorylation of protein substrates. However, this does not result in an increase in hormone secretion rate unless the cytosolic Ca^{2+} concentration is also increased. Thus, in these cells there is a hierarchy of second messengers and cAMP appears to be less important than Ca^{2+}. A third type of relationship is found in blood platelets where a rise in the concentration of free Ca^{2+} leads to a change in shape, aggregation and the exocytosis of secretory granules. In this tissue, however, cAMP directly antagonizes the effect of Ca^{2+} and inhibits these processes. Here, therefore, the two messenger systems work in direct opposition. Thus, cells are often equipped with several second messenger systems and these can interact in different ways to ensure that an appropriate response is made to each particular stimulus.

Further reading

Dual control of adenylate cyclase by G-proteins

Asano, T., Katada, T., Gilman, A.G. and Ross, E.M. (1984). Activation of the inhibitory GTP-binding protein of adenylate cyclase G_i, by β-adrenergic receptors in reconstituted phospholipid vesicles. *J. Biol. Chem.*, **259**, 9351–4.

Bokoch, G.H. (1987). The presence of free G-protein β/γ subunits in human neutrophils results in suppression of adenylate cyclase activity. *J. Biol. Chem.*, **262**, 589–94.

Cerione, R.A., Sibley, D.R., Codina, J., Benovic, J.L. *et al.* (1984). Reconstitution of a hormone sensitive adenylate cyclase system. *J. Biol. Chem.*, **259**, 9979–82.

Gilman, A.G. (1984). G-proteins and dual control of adenylate cyclase. *Cell*, **36**, 577–9.

Helmreich, E.J.M. and Pfeuffer, T. (1985). Regulation of signal transduction by β-adrenergic hormone receptors. *Trends Pharmacol. Sci.*, **6**, 438–43.

Heyworth, C., Whetton, A., Kinsella, A.R. and Houslay, M.D. (1984). The phorbol ester, TPA inhibits glucagon stimulated adenylate cyclase activity. *FEBS Lett.*, **170**, 38–42.

Houslay, M. (1983). Dual control of adenylate cyclase. *Nature*, **303**, 133.

Jakobs, K.H., Aktories, K. and Schultz, G. (1984). Mechanism of pertussis toxin action on the adenylate cyclase system. *Eur. J. Biochem.*, **140**, 177–81.

Jard, S., Cantau, B. and Jakobs, K.H. (1981). Angiotensin II and α-adrenergic agonists inhibit rat liver adenylate cyclase. *J. Biol. Chem.*, **256**, 2603–6.

Katada, T., Kusakabe, K., Oinuma, M. and Ui, M. (1987). A novel mechanism for the inhibition of adenylate cyclase via inhibitory GTP-binding proteins. *J. Biol. Chem.*, **262**, 11897–900.

Levitzki, A. (1986). β-Adrenergic receptors and their mode of coupling to adenylate cyclase. *Physiol. Rev.*, **66**, 819–54.

Levitzki, A. (1987). Regulation of hormone sensitive adenylate cyclase. *Trends Pharmacol. Sci.*, **8**, 299–303.

Levitzki, A. (1987). Regulation of adenylate cyclase by hormones and G-proteins. *FEBS Lett.*, **211**, 113–18.

Levitzki, A. (1988). Transmembrane signalling to adenylate cyclase in mammalian cells and Sacharomyces cerevisiae. *Trends Biochem. Sci.*, **13**, 298–303.

Levitzki, A. (1988). From epinephrine to cAMP. *Science*, **241**, 800–6.

Limbird, E.E. (1981). Activation and attenuation of adenylate cyclase. *Biochem. J.*, **195**, 1–13.

Limbird, L.E. (1988). Receptors linked to inhibition of adenylate cyclase – additional signalling mechanisms. *FASEB J.*, **2**, 2686–95.

Pfeuffer, E., Dreker, R.M., Metzger, H. and Pfeuffer, T. (1985). Catalytic unit of adenylate cyclase: Purification and identification by affinity crosslinking. *Proc. Natl. Acad. Sci. USA.*, **82**, 3086–90.

Schramm, M. and Selinger, Z. (1984). Message transmission: receptor controlled adenylate cyclase system. *Science*, **225**, 1350–6.

Swillens, S. and Dumont, J.E. (1980). A unifying model of current concepts and data on adenylate cyclase activation by β-adrenergic agonists. *Life Sciences*, **27**, 1013–28.

Control of adenylate cyclase by Ca^{2+}

Bradham, L.S. and Cheung, W.Y. (1980). Calmodulin-dependent adenylate cyclase. *Calcium and Cell Function*, (Cheung W.Y. ed; Academic Press, New York), **1**, 109–26.

MacNeil, S., Walker, S.W., Senior, H.J. *et al.* (1984). Calmodulin activation of adenylate cyclase in the mouse B16 melanoma. *Biochem. J.*, **224**, 453–60.

MacNeil, S., Lakey, T. and Tomlinson, S. (1985). Calmodulin regulation of adenylate cyclase activity. *Cell Calcium*, **6**, 213–26.

Thams, P., Capito, K. and Hedeskov, C.J. (1982). Differential effects of Ca^{2+}-calmodulin on adenylate cyclase activity in mouse and rat pancreatic islets. *Biochem. J.*, **206**, 97–102.

Valverde, I., Garcia-Morales, P., Ghiglione, M. and Malaisse, W.J. (1983). The stimulus–secretion coupling of glucose-induced insulin release. LIII. Calcium-dependency of the cyclic AMP response to nutrient secretagogues. *Horm. Metabol. Res.*, **15**, 62–8.

cAMP degradation

Cheung, W.Y., Lynch, T.J., Wallace, R.W. and Tallant, E.A. (1981). cAMP renders Ca^{2+}-dependent phosphodiesterase refractory to inhibition by a calmodulin-binding protein (calcineurin). *J. Biol. Chem.*, **256**, 4439–43.

Houslay, M.D. (1985). Insulin, glucagon and the receptor-mediated control of cyclic AMP concentrations in liver. *Biochem. Soc. Trans.*, **14**, 183–93.

Lin, M.L. and Cheung, W.Y. (1980). Ca^{2+}-dependent cyclic nucleotide phosphodiesterase. *Calcium and Cell function*, (Cheung, W.Y. ed; Academic Press, New York), **1**, 79–107.

Marchmont, R.J., Ayad, S.F. and Houslay, M.D. (1981). Purification and properties of the insulin-stimulated cyclic AMP phosphodiesterase from rat liver plasma membranes. *Biochem. J.*, **195**, 645–52.

Sams, D.J. and Montague, W. (1972). The role of adenosine 3′:5′-cyclic monophosphate in the regulation of insulin release. *Biochem. J.*, **129**, 945–52.

Sugden, M.C. and Ashcroft, S.J.H. (1981). Cyclic nucleotide phosphodiesterase of rat pancreatic islets. *Biochem. J.*, **197**, 459–64.

Sutherland, E.W. and Rall, T.W. (1958). Fractionation and characterization of a cyclic ribonucleotide formed by tissue particles. *J. Biol. Chem.*, **232**, 1077–91.

Mechanism of action of cAMP

Buchler, W. (1988). The catalytic subunit of protein kinase A is essential for cAMP regulation of gene expression. *FEBS Lett.*, **228**, 239–52.

Clegg, C.H., Correll, L.A., Cadd, G.G. and McKnight, G.S. (1987). Inhibition of intracellular cAMP-dependent protein kinase using mutant genes of the regulatory type I subunit. *J. Biol. Chem.*, **262**, 13111–19.

Day, R.N., Walder, J.A. and Maurer, R.A. (1989). A protein kinase inhibitor gene reduces both basal and multihormone-stimulated prolactin gene transcription. *J. Biol. Chem.*, **264**, 431–6.

Doskeland, S.O. and Ogreid, D. (1980). Binding proteins for cyclic AMP in mammalian tissues. *Int. J. Biochem.*, **13**, 1–19.

Ekanger, R., Sand, T.E., Ogreid, D. *et al.* (1985). The separate estimation of cAMP intracellularly bound to the regulatory subunits of protein kinase I and II in glucagon-stimulated rat hepatocytes. *J. Biol. Chem.*, **260**, 3393–401.

Gold, G.H. and Nakamura, T. (1987). Cyclic nucleotide-gated conductances: a new class of ion channels mediates visual and olfactory transduction. *Trends Pharmacol. Sci.*, **8**, 312–16.

Grove, R., Price, D.J., Goodman, H.M. and Avruch, J. (1987). Recombinant fragment of protein kinase inhibitor blocks cyclic AMP-dependent gene transcription. *Science*, **283**, 530-3.

Hoppe, J. (1985). cAMP-dependent protein kinases: conformational changes during activation. *Trends, Biochem. Sci.*, **10**, 29-31.

Khatra, B., Printz, R., Cobb, C. and Corbin, J.D. (1985). Regulatory subunit of cAMP-dependent protein kinase inhibits phosphoprotein phosphatase. *Biochem. Biophys. Res. Comm.*, **130**, 567-73.

Kondrashin, A.A. (1985). Cyclic AMP and regulation of gene expression. *Trends Biochem. Sci.*, **10**, 97-8.

Livesey, S.A., Collier, G. and Zajac, J.D. *et al.* (1984). Characteristics of selective activation of cyclic AMP-dependent protein kinase isoenzymes by calcitonin and prostaglandin E_2 in human breast cancer cells. *Biochem. J.*, **224**, 361-70.

Ogreid, D. and Doskeland, S.O. (1982). Activation of protein kinase isoenzymes under near physiological conditions. *FEBS Lett.*, **150**, 161-6.

Wynshaw-Boris, A., Lugo, T.G., Short, J.M. *et al.* (1984). Identification of a cAMP regulatory region in the gene for rat cytosolic phosphoenolpyruvate carboxy-kinase. *J. Biol. Chem.*, **259**, 12161-9.

Guanylate cyclase and cGMP

Bentley, J.K., Tubb, D.J. and Garbers, D.L. (1986). Receptor mediated activation of spermatozoan guanylate cyclase. *J. Biol. Chem.*, **261**, 14859-62.

Cantin, M. and Genest, J. (1986). The heart as an endocrine gland. *Sci. Am.*, **255**, 62-7.

Chinkers, M. Garbers, D.L., Chang, M-S. *et al.*. (1989). A membrane form of guanylate cyclase is an atrial natriwetic peptide receptor. *Nature*, **338**, 78-83.

Goldberg, N.D. and Haddox, M.K. (1977). cGMP metabolism and involvement in biological regulation. *Ann. Rev. Biochem.*, **46**, 823-96.

Houslay, M. (1985). Renaissance for cyclic GMP? *Trends Biochem. Sci.*, **10**, 465-7.

Hurley, J.B. (1987). Molecular properties of the cGMP cascade of vertebrate photoreceptors. *Ann. Rev. Physiol.*, **49**, 793-812.

Inagami, T. (1989). Atrial natriuretic factor. *J. Biol. Chem.*, **264**, 3043-6.

Kamisaki, Y., Saheki, S., Nakene, M. *et al.* (1986). Soluble guanylate cyclase from rat lung exists as a heterodimer. *J. Biol. Chem.*, **261**, 7236-41.

Kaupp, U.W. and Koch, K.W. (1986). Mechanism of photoreception in vertebrate vision. *Trends Biochem. Sci.*, **11**, 43-7.

Kramer, H.J. (1988) Atrial natriuretic hormones. *Gen. Pharmacol.*, **19**, 747-54.

Kuno, T., Andressen, J.W., Kamisaki, Y. *et al.* (1986). Co-purification of an atrial natriuretic factor receptor and particulate guanylate cyclase from rat lung. *J. Biol. Chem.*, **261**, 5817-23.

Lamb, T.D. (1986). Transduction in vertebrate photoreceptors: the roles of cyclic GMP and calcium. *Trends Neurosci.*, **10**, 224-8.

Lecki, B. (1987). How the heart rules the kidneys. *Nature*, **326**, 644-5.

Paul, A.K., Marala, R.B., Jaiswal, R.K. and Sharma, R.K. (1987). Co-existence of guanylate cyclase and atrial natriuretic factor receptor in a 180-kD protein. *Science*, **235**, 1224-6.

5 Inositol phospholipid turnover in cell calcium homeostasis

Ca^{2+} ions and cell physiology

It has been evident for more than a century that Ca^{2+} ions are vitally important components of the medium which bathes cells. In 1883 Ringer demonstrated that Ca^{2+} was necessary for contraction of heart muscle and, in the intervening years, it has become clear that this metal occupies a unique position among the plethora of ions and compounds upon which cells depend. Most of the Ca^{2+} present in mammals is found in bone and only about 5% is available to the remaining tissues. Nevertheless, this Ca^{2+} is necessary for maintenance of cell structure (by stabilization of membranes and regulation of the cytoskeleton), control of ion permeability (especially Na^+ and K^+), and regulation of mobility and contraction, plus an enormous range of other functions which combine to control the state of cell activation. Indeed, it has been suggested that Ca^{2+} is used as a coupling factor for the initiation of particular responses in almost all types of differentiated cell. The reasons for the selection of Ca^{2+} as the mediator of such diverse cellular functions presumably reflects certain aspects of the unique chemistry of this ion. In particular, it is likely that the combination of the ionic radius of the hydrated ion coupled with its attendant charge density has allowed the evolutionary development of proteins which can readily discriminate Ca^{2+} from all other common di- and mono-valent cations. In conjunction with these proteins, cells have also developed sophisticated Ca^{2+}-buffering systems which involve Ca^{2+} transporters located both at the plasma membrane and on the membranes of intracellular organelles. This enables them to regulate, very precisely, the concentration of Ca^{2+} in the cytosol, and it is this capacity which allows Ca^{2+} to function effectively as a second messenger. The total amount of Ca^{2+} in the cytosol is high (in the millimolar range) but most of this is bound to membranes and cytosolic proteins such that the concentration of free (i.e. unbound, chemically reactive) Ca^{2+} is much lower and is usually in the range $0.1–1 \mu M$ under resting conditions. This means that there is a very large gradient favouring Ca^{2+} influx into the cell since the free Ca^{2+} concentration of the extracellular fluids is about $1 \, mM$ in most vertebrates. This, in turn, suggests an obvious mechanism by which a cell can alter the cytosolic free Ca^{2+} concentration. If a mechanism exists to regulate the rate of Ca^{2+} entry through the plasma membrane, the cell can readily employ this pathway as the means to transduce a hormone signal. In practice, cells have developed a number of variations on this theme and examples of Ca^{2+} channels which are directly controlled by hormones and channels which respond to agonist-induced changes in the membrane potential have been described. It was the desire to understand how these various Ca^{2+} 'gates' are regulated that has led to the recent unravelling of the details of a signalling system which is present in many cells and which links hormone receptors to Ca^{2+}-dependent responses.

Inositol lipids in signal transduction

Animal cells contain a variety of phospholipid species in their membranes, including a class of molecules which have *myo*-inositol (Fig. 5.1) or a derivative as the polar head group. These lipids comprise only approximately 5–10% of the total phospholipid pool but represent a metabolically-active set of molecules. As early as the 1950s it was demonstrated that stimulation of cholinergic receptors in pancreas slices leads to turnover of the phosphate groups in phosphatidylinositol (1-(3-sn-phosphatidylinositol))(PI) and more recently this effect has been attributed to a net breakdown of the lipid. This response forms the basis of a signal transduction system which results in mobilization of Ca^{2+} from intracellular storage sites.

Inositol-containing phospholipids

Animal cells contain three distinct species of phospholipid which contain *myo*-inositol, the most abundant of which is 1-(3-sn-phosphatidyl)-D-*myo*-inositol or phosphatidylinositol (PI). This molecule has the usual glycerophospholipid structure with two long-chain fatty acids attached to positions 1 and 2 of the glycerol backbone but has *myo*-inositol linked via a phosphate group to carbon atom 3 of glycerol. PI can be sequentially phosphorylated by specific kinase enzymes leading to the production of phosphatidylinositol-4-phosphate (PIP) and phosphatidylinositol-4, 5-bisphosphate (PIP_2) respectively, by addition of phosphate groups to hydroxyl residues in the inositol ring

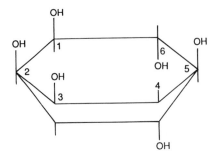

Fig. 5.1 The structure of *myo*-inositol

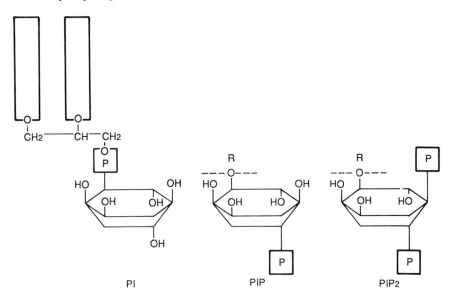

Fig. 5.2 The structures of phosphatidylinositol (PI), phosphatidylinositol-4-phosphate (PIP), phosphatidylinositol-4, 5-bisphosphate (PIP$_2$)

(Fig. 5.2). These two species are present in much smaller amounts than PI and together make up only 10% of the total inositol lipid pool. Many cells contain phosphomonoesterases which can degrade PIP$_2$ and PIP to regenerate PI and thus the amount of these lipids depends upon their relative rates of synthesis from, and degradation to, PI. PIP and PIP$_2$ have become known as polyphosphoinositides (PPI) to denote the presence of the extra phosphate groups in their structures. Precise determination of the relative amounts of the three types of inositol-containing phospholipid in membranes has proved to be quite difficult as the PPI are extremely susceptible to degradation and are often rapidly hydrolysed during tissue preparation. They require acidified chloroform/methanol mixtures for extraction, reflecting the highly anionic nature of their head groups which interact strongly with membrane proteins.

Biosynthesis of inositol phospholipids

PI is synthesized by the addition of *myo*-inositol to CDP-diacylglycerol which is formed by the condensation of CTP and phosphatidic acid in a reaction catalysed by the enzyme CTP-phosphatidic acid cytidyltransferase. *Myo*-inositol can be taken into the cell from the extracellular medium since many cells possess a specific transporter mechanism which facilitates *myo*-inositol uptake. This is an energy-dependent process and can result in the active accumulation of inositol against a concentration gradient. An alternative pathway also exists whereby inositol can be synthesized directly from glucose, but this appears to be a quantitatively less important mechanism since the

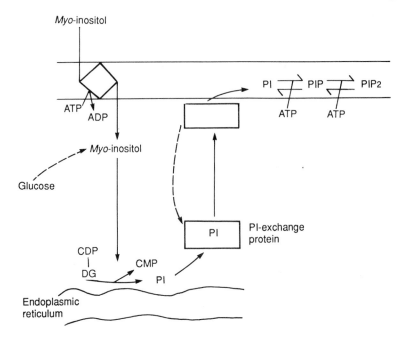

Fig. 5.3 Synthesis of inositol lipids in cells

activity of the rate-limiting enzyme in this pathway (inositol-1-phosphate synthase) is usually low. Biosynthesis of PI occurs at the endoplasmic reticulum (ER) and is catalysed by PI synthase. The PI is then transported from the ER to the plasma membrane by an exchange protein whose characteristics have not been well defined. The phosphorylation of PI to yield PIP is catalysed by a Mg^{2+}-requiring kinase which uses ATP as the phosphoryl donor and takes place at the plasma membrane (Fig. 5.3). PIP_2 formation also occurs at this site but may be catalysed by a separate kinase from that involved in the generation of PIP. It is thought that $PIP + PIP_2$ are predominantly localized to the inner leaflet of the plasma membrane as the enzymes involved in their synthesis appear to be restricted to this location.

PI kinase

PI kinase is a Mg^{2+}-dependent enzyme which is inhibited by Ca^{2+} and by the aminoglycoside antibiotic neomycin. It is believed to be a plasma-membrane associated enzyme, although similar activities have also been found in Golgi and secretory granule membrane preparations. The enzyme has not yet been purified and so little is known of its physicochemical properties. PIP_2 synthesis appears to be catalysed by a separate enzyme since the relative rates of PIP and PIP_2 formation can be altered by manipulation of the ionic conditions under which assays are performed. This enzyme is also Mg^{2+}-

dependent and may be subject to feedback inhibition by its product, PIP_2. In the brain, this protein seems to be present in the cytosol where it has been identified as a 45-K molecular weight component, but in liver homogenates its activity is 50 times greater in the plasma membrane fraction than in other cellular compartments.

PPI phosphomonoesterase

Phosphomonoesterase activities exist which can degrade PIP_2 to PIP and PIP to PI and are Ca^{2+}-activated enzymes which are localized to both the plasma membrane and cytosol fractions. It is not clear whether the same enzyme catalyses both reactions in the cell, although human red-blood cells contain an activity which is apparently specific for PIP_2 and does not hydrolyse PIP.

Polyphosphoinositide turnover and receptor activation

As indicated above, it has been known for a long time that stimulation of cells with particular hormones leads to the breakdown of PI. Following the chemical characterization of the PPI lipid fraction in the early 1960s, it soon became apparent that PPI turnover can also occur as a result of receptor-binding events in both neural and non-neural tissues. This was demonstrated by monitoring changes in the radioactivity associated with these lipids after pre-labelling cells with ^{32}P and by measuring the generation of their breakdown products, which are water-soluble inositol phosphates. On the basis of these types of study it was proposed that the central event which is controlled by receptor activation is the degradation of PPI by a phosphodiesterase enzyme which cleaves the lipid to yield an appropriate inositol phosphate and diacylglycerol. This reaction has now been characterized in a wide range of tissues and a host of different receptors are able to stimulate it (Table 5.1). The key point which has emerged from an examination of receptor-mediated PPI turnover is that it occurs in tissues which have responses which are dependent on Ca^{2+}. These include enzyme and hormone secretion (e.g. exocrine and endocrine pancreas), smooth muscle contraction (e.g. iris) and glycogen breakdown (liver). Originally the idea had been advanced that somehow the breakdown of PI was related to the Ca^{2+}-dependent responses in these tissues but this proposal was superseded by the demonstration that PPI breakdown is the primary response. It remains unclear whether all three of the inositol-containing phospholipids are broken down directly in stimulated cells, or whether the 'breakdown' of PIP and PI occurs as a result of the conversion of PIP to PIP_2 and PI to PIP, to compensate for an initial hydrolysis of PIP_2. It seems likely that this must occur to some extent, otherwise the PIP_2 pool would rapidly become depleted, rendering the cell refractory to further stimulation. However, the extent of PI loss following hormone stimulation is often much greater than would be necessary to simply replenish the PPI pool. Therefore, it is probable that direct

Table 5.1   Receptors coupled to phosphatidylinositol-4, 5-bisphosphate hydrolysis

Receptor	Tissue	Response
Muscarinic	parotid	fluid secretion
	exocrine pancreas	enzyme secretion
	endocrine pancreas	hormone secretion
Histamine (H_1)	adrenal	catecholamine secretion
Purinergic (P_2)	liver	glycogenolysis
Adrenergic (α_i)	liver	glycogenolysis
	smooth muscle	vasoconstriction
Thrombin	platelet	aggregation
Thryotropin-releasing hormone	pituitary	prolactin secretion
Vasopressin (V_1)	liver	glycogenolysis
	smooth muscle	vasoconstriction
Serotonin	insect salivary gland	fluid secretion
	platelets	secretion
Bombesin	endocrine pancreas	hormone secretion

PI hydrolysis does occur in some cells and although it is now clear that this does not directly alter the cell's Ca^{2+} status (see below), it does lead to the production of diacylglycerol in the membrane. This may be an important signalling factor, since diacylglycerol can activate a specific protein kinase in the cell (protein kinase C) and can, therefore, regulate protein phosphorylation independently of any change in either cAMP or cytosolic free Ca^{2+}.

Phospholipase C

A phosphoinositide-specific phospholipase C is the phosphodiesterase enzyme responsible for catalyzing PPI hydrolysis, and it has been the subject of considerable recent scrutiny in an attempt to define its characteristics, properties and mode of activation.

It has been assumed for many years that phospholipase C (PLC) is primarily a cytosolic enzyme, since its activity can easily be measured in soluble extracts of cells. However, it is also clear that the substrates are located in cellular membranes which must imply that the enzyme can, and does, interact with membranes. Indeed, it has recently been shown that PPI hydrolysis can be initiated under appropriate conditions in membrane preparations isolated from liver, pituitary and pancreas which are devoid of cytosolic components (Fig. 5.4). This suggests that these membranes must contain an endogenous PLC. The likelihood is, that both cytosolic- and membrane-bound forms of the enzyme occur but that the latter is more important for initiating receptor-activated PPI hydrolysis. It could be envisaged that hormones might induce a

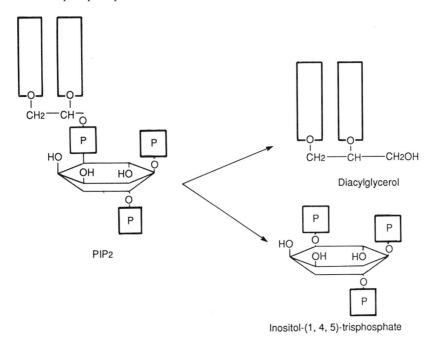

Fig. 5.4 Hydrolysis of PIP_2 by phospholipase C

redistribution of the enzyme from the cytosol to the plasma membrane in order to promote PPI breakdown but experimental evidence for this idea is lacking. Furthermore, the time scale within which hormones can initiate PPI hydrolysis (less than 2 seconds in insect salivary gland and insulinoma cells!) is not entirely consistent with such a mechanism (but see below).

Immunological evidence has suggested that at least four types of PLC may be present in cells, although the precise relationship between these enzymes has not been established. The relative amounts of the different forms seems to vary from tissue to tissue but it is not yet clear how this is related to their individual functional roles in each tissue.

There has been some dispute concerning the Ca^{2+} sensitivity of PLC and about its substrate specificity. As described above, in the intact cell the enzyme preferentially hydrolyses PIP_2 to generate $I-(1,4,5)P_3$ and diacylglycerol (Fig. 5.4). However, the same enzyme also hydrolyses PI and PIP in tissue extracts, raising doubts abouts its specificity. The solution to this dilemma seems to relate to the prevailing Ca^{2+} concentration used in the assay systems. Detailed studies of the Ca^{2+} dependence of PLC have revealed that the enzyme is not active if the free Ca^{2+} concentration falls into the nanomolar range, but that at free Ca^{2+} concentrations close to those found in resting cells ($0.1-1\mu M$) PLC preferentially degrades PIP_2. This clearly differentiates the hormone-sensitive enzyme from a similar activity present in lysosomes which does not require Ca^{2+} for activity. If the Ca^{2+} concentration is raised further, PI hydrolysis also begins to occur, although the exact Ca^{2+} dependency of this latter response remains uncertain. It appears that a Ca^{2+}-concentration of between $10\mu M$ and 1 mM is necessary to allow hydrolysis of PI, and it is possible that cells may regulate the rate of PI breakdown according to the cytosolic concentration of Ca^{2+}. Thus, under resting conditions activation of a hormone receptor will cause preferential breakdown of PIP_2 which will, in turn, lead to mobilization of stored Ca^{2+} (see below). The rise in Ca^{2+} concentration may then facilitate PI hydrolysis and a greater production of diacylglycerol.

It is important to note however that PLC is *not* directly activated by Ca^{2+}. Simply raising the cytosolic Ca^{2+} concentration does not lead to either PIP_2 or PI breakdown in most cells, which only occurs when a hormone or drug binds to an appropriate receptor on the cell surface. This is an important concept since it underlies the function of PIP_2 hydrolysis as a messenger system for Ca^{2+} mobilization. However, as with all rules, there are exceptions to this generalization, as for example, in pancreatic B-cells and parts of the brain. In these tissues, an increase in Ca^{2+}-influx rate can directly stimulate PLC activity and, therefore, initiates PIP_2 breakdown. The significance of this response may lie in the resultant production of diacylglycerol which can synergistically interact with Ca^{2+} in, for example, the stimulation of insulin secretion. In general, however, Ca^{2+}-influx does not directly increase the activity of PLC.

The reaction mechanism by which PIP_2 hydrolysis proceeds involves cleavage of the bond between the phosphate group at position 1 of the inositol ring and the glycerol backbone. The hydrolytic mechanism consumes a hydroxyl ion during the reaction which can be provided either by water or by inositol itself. In the former case the reaction product is I-(1,4,5)P_3 but if a hydroxyl from the inositol ring is employed, the product is a cyclized derivative, inositol 1:2-(cyclic)-4,5-trisphosphate (Fig. 5.5). The precise ratio in which these two products are generated in the cell is still a matter of dispute and it is not certain whether they represent separate messenger molecules.

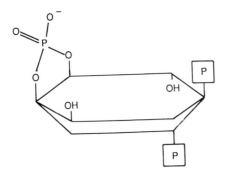

Fig. 5.5   Inositol 1:2-(cyclic)-4,5-trisphosphate

Regulation of phospholipase C (PLC) activity

The activity of PLC can be regulated by the presence of other, non-substrate, lipids in the plasma membrane since the disposition of the component lipids is crucial to the ability of the enzyme to hydrolyse its substrate. Lipid derivatives which do not insert into membrane bilayers are ineffective as either substrates or inhibitors of PLC but the presence of large amounts of phosphatidylcholine in a membrane serves to markedly inhibit the enzyme. This effect can be overcome by other lipids such as phosphatidylserine or diacylglycerol and it may be that one of the mechanisms by which hormone receptors control PLC activity is by changing the microenvironment in which the substrate lipids reside, such that they are more readily accessible to the enzyme. If this is the mechanism by which enzyme activation is effected, then PPI turnover is a second messenger system which is controlled in a different way from the cAMP system (see Chapter 4). In the case of cAMP, its concentration is raised by a hormone-induced increase in adenylate cyclase activity by a mechanism which leads to a true increase in the intrinsic activity of the enzyme. In the case of PLC however, activation by altering the substrate conformation would lead to a more rapid reaction velocity without any intrinsic change in the activity of the enzyme. The end result in each case is the same, however, being an increased concentration of a second messenger molecule.

Guanine-nucleotides as regulators of PLC

The discovery that adenylate cylase activity is regulated by two particular transducer proteins $(G_s + G_i)$ which transmit the hormone 'signal' from receptor to adenylate cyclase has prompted the search for similar molecules which could act as regulators of PLC. Evidence for their existence was first provided by results which revealed that in neutrophils and mast cells pertussis toxin causes ADP-ribosylation of a specific membrane protein and, at the same time, leads to inhibition of receptor activated PPI breakdown. Since the first pertussis toxin substrate to be discovered was G_i, and the ADP-ribosylated protein found in these membranes had a molecular weight of 41 K (similar to that of G_i), it was suggested that G_i could be important in control of PLC. However, in other cells which also contain G_i (e.g. liver, pancreatic islets) pertussis toxin does not affect receptor-stimulated PPI turnover. Furthermore, it is now clear that a family of G-proteins exist (see Chapter 3), many of which are similar in size to G_i but which can be distinguished immunologically. It is possible therefore that neutrophils and mast cells contain a second pertussis toxin sensitive G-protein, which is distinct from G_i, but which may somehow be linked to PLC. Since in other cells PPI turnover is not affected by pertussis toxin, this suggests that either different cells must employ different transducer proteins in the regulation of PLC or that the apparent link between ADP-ribosylation of G_i and PLC activity in neutrophils is simply coincidental.

More direct evidence for the involvement of a G-protein in PLC activation

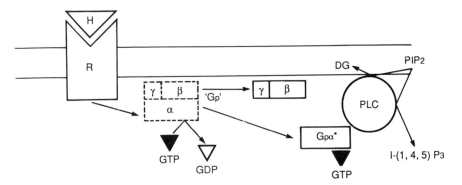

Fig. 5.6 Possible mode of activation of phospholipase C by a G-protein coupled to stimulatory receptors

has come from observations demonstrating that GTP and its non-hydrolysable analogues (e.g. GTP-γ-S) can directly stimulate PPI breakdown and inositol phosphate production in membrane preparations isolated from liver, smooth muscle, pituitary, pancreas and an increasingly long list of other cells. In some cases this occurs in the complete absence of a hormone, whereas, in others, the guanine-nucleotide potentiates the resonse to a receptor agonist (Fig. 5.6). In all cases, the consensus suggests that a G-protein intervenes in the coupling mechanism which links the cell-surface receptor to PLC. The biochemical nature of this protein remains to be determined but it has been tentatively named G_p to denote its interaction with, or effects on PLC activity.

One interesting possibility is that G_p may be a member of the *ras* proto-oncogene family of proteins which are believed to be GTP-binding proteins. These molecules are all of about 21-K molecular weight and it has been shown that over-expression of one of them in genetically-manipulated fibroblasts can lead to enhanced PPI turnover in response to the peptide growth factor bombesin. In addition, other cells transformed with oncogenic variants of *ras* genes exhibit rates of PPI breakdown and levels of inositol phosphates which are higher than those of control cells. This evidence, therefore, implicates the 21 K protein in control of PLC activity but it is still too early to draw definite conclusions as to whether this represents the normal coupling protein for all receptors.

The mechanism by which G_p might couple to PLC is also in doubt since any direct interaction would not accord with the possibility that the rate of PPI hydrolysis is regulated by the lipid microenvironment surrounding the substrate. It is conceivable that G_p could control the accessibility of PIP_2 in the plasma membrane but mechanistically, it is easier to envisage a direct regulation of PLC activity. This possibility has been explored in platelets where a cytosolic PLC which is sensitive to guanine nucleotides has now been identified. This enzyme hydrolyses PIP_2 preferentially and its activity can be increased as much as 25-fold by GTP-γ-S. The effect of GTP-γ-S requires the presence of another cytosolic component and it is suggested that both the enzyme and its coupling protein are present in the soluble fraction of the cell, but that binding of GTP promotes the association of the complex with the plasma membrane. This model raises the question as to how a cell-surface receptor can be coupled to a cytosolic G-protein. It could be speculated that the G-protein is released from the membrane on receptor binding (as has been suggested for G_s) but more information is needed before the biochemistry of PLC activation will be fully understood.

Inositol phosphate phosphomonoesterase

The products released as a result of PLC-catalysed hydrolysis of the inositol-containing phospholipids are I-(1,4,5)P_3, I-(1,4)P_2 and inositol-1-phosphate, respectively (Fig. 5.7). At least one of these compounds has physiological activity [I-(1,4,5)P_3] and therefore its metabolism is an essential part of the second messenger system, since this can 'switch-off' the hormone signal. Degradation of all of these molecules can proceed by sequential dephosphorylation and cells are equipped with phosphomonoesterase enzymes to achieve this. One result of the action of these enzymes is that free *myo*-inositol can be regenerated which is then available for use in the re-synthesis of PI to complete the 'PI cycle'. The phosphatases involved in inositol phosphate metabolism have been characterized and show considerable substrate specificity. IP_3 phosphatase specifically hydrolyses the phosphomonoester bond at

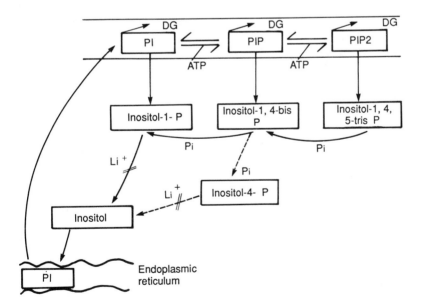

Fig. 5.7 The inositol lipid cycle

position 5 of the inositol ring to convert I-$(1,4,5)P_3$ to I-$(1,4)P_2$. The enzyme has been located in both cytosol and membrane preparations and requires Mg^{2+} for activity. The activity of the enzyme may be regulated by phosphorylation since it is a substrate for protein kinase C, and phosphorylation induces a 4-fold increase in catalytic activity. This could form part of a feedback mechanism controlling cell activation since protein kinase C becomes activated in the presence of the diacylglycerol which is generated by PIP_2 break down. If this then leads to more rapid IP_3 degradation, the result will be a termination of the hormone signal.

I-$(1,4)P_2$ can be dephosphorylated by two enzymes: an inositol-1-phosphatase or an inositol-4-phosphatase. The relative activities of these two enzymes may vary between species and from tissue to tissue, since in liver and rat brain both enzymes are present, whereas in calf brain the 1-phosphatase predominates. This latter enzyme is inhibited by Li^+ ions which also reduce the activity of the final enzyme in the sequence, inositol monophosphate phosphomonoesterase. In intact cells, as little as 0.1 mM Li^+ can reduce the activity of the monophosphatase, an observation which has been exploited in investigations of the inositol lipid signalling system by incubating cells in media containing Li^+ to effectively block the re-cycling of inositol. This has the effect of causing a large build-up of IP in the cell and thus provides a convenient experimental tool to study inositol polyphosphate generation. If cells are pre-labelled with ^{3}H-*myo*-inositol this becomes incorporated into the inositol phospholipids and their hydrolysis can then be readily followed by monitoring the production of ^{3}H-inositol phosphates following agonist treatment. The different inositol phosphate species can be resolved on anion-exchange HPLC columns and their relationships studied. This technique has now found application in a wide variety of cells which exhibit receptor-linked PIP_2 breakdown.

Inositol trisphosphate as a second messenger

The demonstration that one of the inositol phosphates produced in the PI cycle can mobilize a pool of intracellular Ca^{2+} was crucial to the development of our understanding of the inositol lipid signalling system. This observation was first made in pancreatic acinar cells which had been made permeable by incubation in a medium containing a low concentration of Ca^{2+}. In the permeable cells, Ca^{2+} could be accumulated into a membrane-bound pool by a mechanism which required the presence of ATP. If IP_3 was then added to the cells, it could readily gain access to the intracellular environment (normally cells are impermeable to highly-charged molecules like IP_3) and cause a rapid release of Ca^{2+}. The K_m for IP_3-induced Ca^{2+} release was about 1 μM which is within the range of concentration that calculations show could be generated from PIP_2 in a stimulated cell, and the intracellular pool which was sensitive to IP_3 appeared to be the same as that mobilized following stimulation of muscarinic acetylcholine receptors on pancreatic cells (Fig. 5.8).

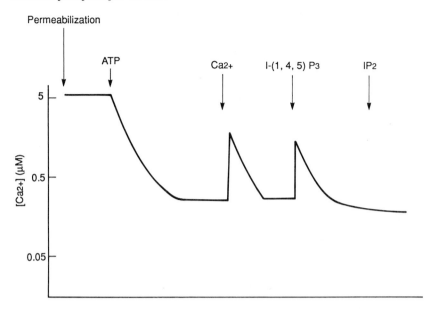

Fig. 5.8 Uptake and release of Ca^{2+} from intracellular pools in permeabilized cells

The specificity of this response was extremely marked, since IP_2 and several other inositol phosphates were virtually inactive at similar concentrations to that at which IP_3 elicited a large release of Ca^{2+}. Thus, IP_3 was established as a prime candidate for the calcium-mobilizing second messenger, and it has now been shown that this molecule can cause intracellular Ca^{2+} release in a wide variety of cell types.

One potential problem which arose with this interpretation was whether a receptor-mediated rise in IP_3 could occur with sufficient rapidity to account for the physiological response. For example, it was established in hepatocytes that the Ca^{2+}-dependent activation of phosphorylase a occurred several seconds prior to any measurable breakdown of PI and that the level of IP_3 was not maximal for as long as 2 min after treatment of liver cells with vasopressin, whereas the maximal rise in cytosolic Ca^{2+} occurred within 10–15 sec. Detailed time-course studies revealed, however, that IP_3 formation begins within 1–2 sec of hormone addition and precedes the rise in cytosolic Ca^{2+} concentration. The second messenger role of IP_3 has now been confirmed by the demonstration that microinjection into photoreceptor cells of *Limulus* (horseshoe crab) or *Xenopus* oocytes lead to Ca^{2+} mobilization and cell activation.

The IP_3-sensitive intracellular Ca^{2+} pool

Since mitochondria contain substantial quantities of labile Ca^{2+}, it was assumed for many years that they represented the location of hormone-sensitive stores. Indeed, treatment of liver cells with 'Ca^{2+}-mobilizing hormones' was found to cause a large reduction in the Ca^{2+} content of subsequently isolated mitochondria, which added weight to this idea. However, when permeable cell preparations became available it was found that a second, lower-capacity but higher-affinity pool of labile Ca^{2+} was also present. This was independent of the mitochondria since it could function despite the presence of mitochondrial inhibitors (e.g. oligomycin, rotenone). This pool was located in a membrane-bound compartment since it could be released by ionophores such as ionomycin and A23187, and it accumulated Ca^{2+} in an ATP-dependent manner. The fraction appears to be part of the endoplasmic reticulum (ER) of the cell, although it is not yet clear whether a specialized region of this structure participates in Ca^{2+} homeostasis or whether the whole of the ER can subserve this function. The identity of this fraction with ER was confirmed by the demonstration that isolated microsomes from rat insulinoma cells can release Ca^{2+} in response to IP$_3$, whereas isolated mitochondria are unaffected. Similar data have also been obtained in other tissues.

The extreme specificity displayed by the ER Ca^{2+}-releasing system suggests that it operates by a receptor mechanism which has a high affinity for IP$_3$. In fact it seems that the phosphates at positions 4 and 5 of the inositol ring must form part of the recognition mechanism since a non-biological isomer of IP$_3$ [I-(2,4,5)P$_3$] is also an effective Ca^{2+}mobilizing agent. Furthermore, the corresponding inositol bisphosphate [I.(4,5)P$_2$] can also induce Ca^{2+} release, although it is less potent than either of the trisphosphate derivatives. Recently, high-affinity IP$_3$ binding sites have been identified in liver, neutrophils and brain using both [^{3}H]-IP$_3$ and [^{32}P]-IP$_3$. The binding of labelled ligand was not inhibited by I-(1,4)P$_2$ or by IP but could be displaced by I-(4,5)P$_2$. In addition there was a close correlation between IP$_3$ binding and Ca^{2+} release in permeable cells, suggesting that the binding sites may represent the physiological IP$_3$ receptors. In the brain the binding of IP$_3$ to its receptor may be regulated by Ca^{2+} in a feedback system since the presence of 0.5 μM Ca^{2+} reduced the binding to only 10% of that found in the absence of Ca^{2+}. Liver and neutrophils did not, however, show a similar Ca^{2+} sensitivity, suggesting differential regulation of IP$_3$ receptor binding in different tissues. The receptor for I(1,4,5)P$_3$ has recently been purified to apparent homogeneity from rat cerebellum and initial characterization of the molecule is now being carried out. The protein has a globular structure as judged by hydrodynamic criteria and comprises a single large subunit of molecular weight 260 K. The receptor shows the correct specificity for inositol phosphate binding suggesting that it probably represents the physiologically relevant molecule. Further studies will now be required to establish how this molecule functions to elicit Ca release from the endoplasmic reticulum.

Mechanisms of IP$_3$-induced Ca-release

Uptake of Ca^{2+} into the ER occurs by a mechanism which requires ATP and can be inhibited by vanadate ions. These ions do not, however, alter the ability

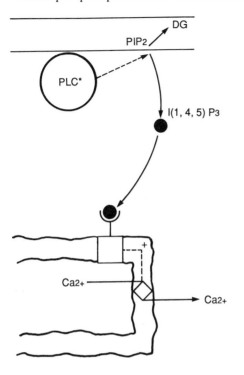

Fig. 5.9 Second messenger function of inositol-(1,4,5)-trisphosphate

of IP_3 to mobilize Ca^{2+} from the ER, suggesting that the mechanism of Ca^{2+} release does not involve any alteration in the functioning of the Ca^{2+}-uptake-system. It has been speculated that IP_3 may directly regulate an ER 'Ca^{2+}-channel' via its receptor and a limited amount of experimental evidence has been provided to support this idea. This does seem an attractive possibility since such a mechanism could readily provide the speed of response necessary for propagation of a hormone signal (Fig. 5.9). However, firm details of the Ca^{2+}-release mechanism must await the results of further experiments. One thing that is clear is that IP_3 can only release a certain fraction of the Ca^{2+} which is accumulated by the ER. This varies somewhat according to the cell type, but generally represents only 25–50% of the Ca^{2+} which can be released by ionophore treatment.

Another facet of the Ca^{2+} release process is that it appears to be transient. If permeable cells are treated with IP_3 a rapid release of Ca^{2+} occurs which is then more slowly re-accumulated by the ER. This may reflect rapid degradation of IP_3 by the permeabilized cells since a second addition of the messenger can elicit further Ca^{2+} mobilization. It is clear that cells rapidly degrade IP_3, which is essential if it is to function effectively as a second messenger. However, it is also evident that IP_3 formation is part of the hormonal amplification system in the cell since calculations suggest that each molecule of IP_3 can release about 20 Ca^{2+} ions from the ER.

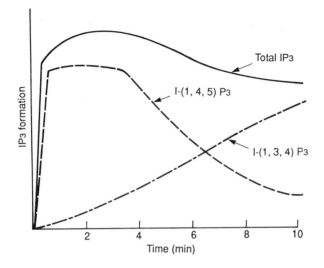

Fig. 5.10   Time course of inositol trisphosphate formation in hormone-stimulated cells

Isomers of IP$_3$

As described above, I-(1,4,5)P$_3$ is believed to represent the second messenger which induces intracellular Ca^{2+} mobilization in stimulated cells, and it is produced very rapidly following hormone stimulation. Indeed, in most cells which display hormone-sensitive PIP$_2$ hydrolysis, IP$_3$ levels remain elevated for many minutes after addition of a receptor agonist. However, it has become clear that the isomeric configuration of the molecule may vary during this time. The original method for isolation of IP$_3$ from cell extracts involved elution of the molecule from an anion exchange resin with ammonium formate. This method was not sufficiently powerful to resolve individual isomers of IP$_3$ but when HPLC methods were employed, such resolution became possible. Studies with parotid gland then provided the first suggestion that the IP$_3$ fraction of stimulated cells may be heterogeneous. Treatment of these cells with carbamoyl choline to activate muscarinic receptors was followed by a rapid generation of I-(1,4,5)P$_3$ as expected, but this was followed after a brief lag period by formation of an IP$_3$ which had slightly different chromatographic properties on anion-exchange resins (Fig. 5.10). This was subsequently identified by chemical analysis as inositol-(1,3,4)-trisphosphate [I-(1,3,4)P$_3$], and it was shown to be the predominant IP$_3$ isomer present in the cells after several minutes of hormone stimulation. Detailed time course analysis has now revealed that hormone stimulation of a wide variety of cells leads to immediate formation of I-(1,4,5)P$_3$ (within 2 sec) which reaches a peak value after 1–2 min and then subsequently declines to a new steady-state concentration which is only slightly above the resting level. In contrast, formation of I-(1,3,4)P$_3$ occurs only after a lag period of ~ 15 sec and its concentration then continues to rise for up to 30 min before it stabilizes at a value which is much higher than the basal level. The net result of this, is that after 10 min or so of hormone stimulation, about 90% of the IP$_3$ present in cells is the (1,3,4)-isomer. This is a surprising finding since I-(1,3,4)P$_3$ has no known second-messenger function. It is not effective at mobilizing intracellular Ca^{2+} since it lacks the phosphate group at position 5 which seems to be essential for this activity. Furthermore, it does not represent an antagonist to the action of I-(1,4,5)P$_3$ since this latter isomer can mobilize Ca^{2+} equally effectively in the presence or absence of I-(1,3,4)P$_3$. At present, the biological function of I-(1,3,4)P$_3$ remains a mystery.

Inositol tetrakisphosphate

The demonstration that stimulated cells accumulate I-(1,3,4)P$_3$ prompted a search for the immediate precursor of this molecule. One possibility which was explored is that two isomers of PIP$_2$ are present in cell membranes, one having a configuration which yields I-(1,4,5)P$_3$ following hydrolysis by PLC (i.e. phosphatidylinositol-4,5-bisphosphate) while the second would be phosphatidylinositol-3,4-bisphosphate. However, no evidence for the existence of this putative lipid precursor could be found, despite extensive analysis of appropriate cell types. This seemed surprising since the amounts of I-(1,3,4)P$_3$ produced in cells, such as parotid and liver after receptor stimulation, would require substantial quantities of precursor lipid if it were generated by this route. A key observation was then made by Batty *et al.* (1985), who observed that muscarinic stimulation of slices of rat cerebral cortex results in the formation of a water-soluble inositol polyphosphate which is even more polar than either of the IP$_3$ isomers. This had previously been missed in other studies since it is only eluted from ion-exchange columns by much more concentrated ammonium formate solutions than are necessary to elute the isomers of IP$_3$. This molecule was subsequently identified as a tetraphosphorylated derivative of *myo*-inositol, having the isomeric configuration, inositol-(1,3,4,5)-tetrakisphosphate [I-(1,3,4,5)P$_4$] (Fig. 5.11). It soon became

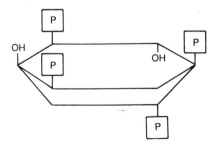

Fig. 5.11 Inositol-(1,3,4,5)-tetrakisphosphate

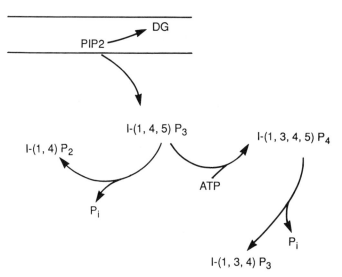

Fig. 5.12   Metabolic routes for inositol-(1,4,5)-trisphosphate in stimulated cells

apparent that this species, like I-(1,4,5)P$_3$, is an early product of hormone-stimulated cells and is detectable several seconds before any measure rise of I-(1,3,4)P$_3$ occurs. Indeed, the kinetics of I-(1,3,4,5)P$_4$ generation are remarkably similar to those of I-(1,4,5)P$_3$. Very surprisingly, I-(1,3,4,5)P$_3$ is not a second messenger for Ca^{2+} mobilization. Presumably the presence of the additional phosphate group at position 3 of the inositol riong is sterically unfavourable for interaction with the ER IP$_3$ receptor.

The demonstration that cells generate a tetra-phosphorylated inositol derivative, the time course of which is similar to that of I-(1,4,5)P$_3$, suggested a possible precursor–product relationship between these two molecules. This was subsequently confirmed when a cytosolic enzyme activity was identified in brain, liver and pancreas homogenates, which catalyses the phosphorylation of I-(1,4,5)P$_3$ to yield I-(1,3,4,5)P$_3$. This enzyme uses ATP as the phosphoryl donor and is completely specific in that it only phosphorylates the *myo*-inositol ring at carbon atom 3. It has a high affinity for I-(1,4,5)P$_3$ (K_m 0.6 μM) and in order to serve as a substrate for the enzyme, the inositol derivative must also have phosphate groups at positions 4 and 5. The enzyme is not however, absolutely specific for I-(1,4,5)P$_3$ since glycerophosphoinositol-4,5-bisphosphate (PIP$_2$ with its fatty acids removed) is also a substrate for the partially-purified enzyme. In contrast, neither I-(1,4)P$_2$ nor I-(1,3,4)P$_3$ can be phosphorylated by the enzyme.

It appears, therefore, that two routes exist for metabolism of I-(1,4,5)P$_3$ in hormone-stimulated cells (Fig. 5.12). It can either be degraded to I-(1,4)P$_2$ by the action of a 5-phosphatase or it can be further phosphorylated under the influence of I-(1,4,5)P$_3$-3-kinase to generate I-(1,3,4,5)P$_4$. Interestingly, both of these routes lead to loss of its second-messenger activity since neither product can mobilize Ca^{2+} stored in the ER. However, formation of IP_4 is an energy-consuming process which utilizes ATP and it seems unlikely that cells would employ this energetically-unfavourable metabolic pathway unless they could drive some direct benefit in so doing.

Recent evidence suggests that the 'direction' of $I-(1,4,5)P_3$ metabolism may be regulated according to the cytosolic free Ca^{2+} concentration in the cell. Studies with rat insulinoma cells have revealed that prior depletion of the ER Ca^{2+} stores is associated with a much larger rise in $I-(1,4,5)P_3$ following muscarinic receptor stimulation than occurs in Ca^{2+} replete cells. At the same time, the production of $I-(1,3,4)P_3$ is correspondingly reduced. Measurements of the activity of $I-(1,4,5)P_3$-3-kinase in homogenates of these cells have shown that the enzyme activity increases when the Ca^{2+} concentration is raised from 0.1 to $10\,\mu M$. Conversely, the activity of the $I-(1,4,5)P_3$-5-phosphatase is not affected by these fluctuations in Ca^{2+} concentration. This Ca^{2+} sensitivity may be mediated by calmodulin, which can activate the enzyme in homogenates of insulinoma and smooth-muscle cells.

It appears, therefore, that in resting cells the basal rate of $I-(1,4,5)P_3$ generation is compensated by the activity of a 5-phosphatase enzyme which degrades the molecule to $I-(1,4)P_2$ and allows the subsequent regeneration and recycling of *myo*-inositol. When, however, a 'Ca^{2+}-mobilizing hormone' binds to its receptor, PLC activity is increased and the rate of $I-(1,4,5)P_3$ generation correspondingly increases. This leads to release of Ca^{2+} from the ER, a rise in cytosolic free Ca^{2+} and activation of Ca^{2+}-dependent processes, such as exocytosis, smooth-muscle contraction or glycogen breakdown according to the cell type. The rise in Ca^{2+} also leads to the calmodulin-dependent activation of $I-(1,4,5)P_3$-3-kinase and therefore to the diversion of IP_3 towards the production of IP_4. Since prevention of this pathway by Ca^{2+} depletion results in reduced $I-(1,3,4)P_3$ levels, this suggests that this second isomer of IP_3 may be directly derived from IP_4. Such a mechanism would readily explain the delayed onset of its production following receptor activation and also its more prolonged rate of synthesis in the cell. In support of this, an enzyme has been identified in red blood cells which can hydrolyse the phosphate group at carbon atom 5 of $I-(1, 3, 4, 5)P_4$ to yield $I-(1,3,4)P_3$. This, therefore, appears to be its route of synthesis. The 5-phosphatase activity may well be the same enzyme that also degrades $I-(1,4,5)P_3$ to $I-(1,4)P_2$, although the V_{max} for the latter reaction is much higher. This may be compensated by a lower K_m for IP_4 and, in the cell, the ratio of the two substrates probably determines their relative rates of hydrolysis.

Functions of inositol tetrakisphosphate

As outlined above, the extra energy expenditure required to produce IP_4 suggests that it may serve a crucial function as an intracellular messenger in its own right. At present, this function is unknown but several recent hints have emerged pointing to a role at the plasma membrane. It is clear that IP_4 has no

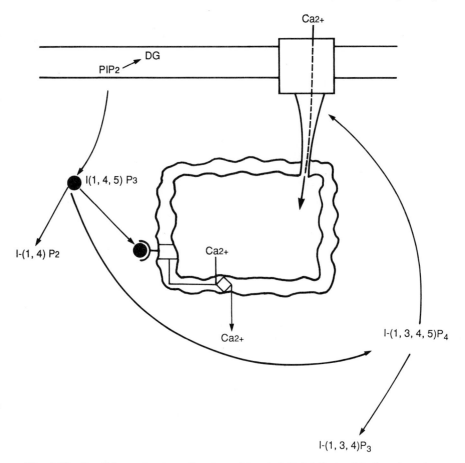

Fig. 5.13 Possible mechanism of action of inositol-(1,3,4,5)-tetrakisphosphate in stimulated cells

effect on intracellular Ca^{2+} stores but it is also clear that the amount of releasable Ca^{2+} present at these sites is only sufficient to maintain an elevated concentration of cytosolic free Ca^{2+} for, at most, a few minutes. More prolonged stimulation ($> 5\,min$) also requires an increase in the rate of Ca^{2+} entry through the plasma membrane from the extracellular fluids which represent an almost infinitely large Ca^{2+} pool relative to that present intracellularly. The possibility that IP_4 may control Ca^{2+} influx rates has been raised by studies using sea-urchin eggs. In these cells, micro-injection of I-$(1,4,5)P_3$ leads to the exocytosis of intracellular secretory granules leading to the formation of a 'fertilization envelope'. This effect is dependent on the presence of extracellular Ca^{2+}, suggesting that it does not simply reflect intracellular mobilization. Indeed, the response is not mimicked by introduction of I-$(2,4,5)P_3$ despite the ability of this compound to cause Ca^{2+} release. One major difference between these two isomers of IP_3 is that the 1,4,5-isomer serves as the preferential substrate for IP_3-3-kinase and can be phosphorylated to I-$(1,3,4,5)P_4$. Surprisingly, however, micro-injection of I-$(1,3,4,5)P_4$ does not lead to the formation of a fertilization envelope in sea-urchin eggs, suggesting that this compound cannot be the sole second messenger responsible for mediating this effect. If however, an isomer of IP_3 is also micro-injected together with IP_4, activation of the eggs follows immediately. This does not seem to reflect a need for intracellular Ca^{2+} mobilization but suggests that IP_3 and IP_4 may have a concerted role to play in regulating the rate of Ca^{2+} inflow through the plasma membrane. The molecular details of this process are not yet understood. One possibility, suggested by Irvine and Moor (1987) based on a model originally proposed by Putney, is that IP_4 somehow promotes the formation of a 'channel' linking the plasma membrane and the ER which facilitates the movement of extracellular Ca^{2+} from outside the cell directly into the ER pool (Fig. 5.13). It is then immediately released from this site into the cytosol due to the presence of IP_3. It is by no means certain how widespread this mechanism may be in other cell types, and it seems probable that a few more surprises may still be in store before this increasingly complex hormonal second messenger system is completely understood.

Other inositol tetrakisphosphate isomers

High-resolution HPLC analysis of the inositol polyphosphates present in bovine adrenal glomerulosa cells stimulated with angiotensin II has now revealed the presence of a second IP_4 isomer in addition to I-$(1,3,4,5)P_4$. Analysis of this molecule has revealed that it lacks a phosphate group at position 5 of the inositol ring, suggesting that it may be derived from I-$(1,3,4)P_3$. Indeed, an enzyme capable of producing the new IP_4 from I-$(1,3,4)P_3$ has been identified in these cells and in liver homogenates. The product appears to be I-$(1,3,4,6)P_4$ based on preliminary chemical analysis but, at present, we know nothing of the possible functions of this molecule.

It is clear from the foregoing that a single receptor-controlled event, namely the hydrolysis of PIP_2, can set in motion an enormous range of intracellular reactions and can lead to the generation of a large number of potential second messenger compounds. It is likely that individual cells may preferentially utilize different members of this family to regulate particular cellular functions, and this opens up a whole new field of possibilities for the integration of cell activation.

Diacylglycerol as a second messenger

In describing the metabolism of the various inositol phosphates produced as a result of activation of PLC, the second product of PIP_2 breakdown, 1,2 diacylglycerol (DG), has been largely overlooked. This is a relatively short-lived molecule in that it remains in the membrane upon hydrolysis of the polar head group, and is rapidly converted to phosphatidic acid by a diacylglycerol kinase. However, its transient existence does not prevent it from playing an important second messenger role in its own right. This lies in the ability of DG to cause the activation of a specific Ca^{2+}-dependent protein kinase, protein kinase C (PKC). The properties of this enzyme are described in more detail in Chapter 8, but it functions independently of the inositol lipid system to alter the phosphorylation state of cellular proteins on specific serine or threonine residues. DG alters the Ca^{2+} sensitivity of PKC such that the enzyme becomes activated at Ca^{2+} concentrations as low as $0.1 \mu M$. This effect seems to require that one of the fatty acids present in DG is unsaturated, and since the DG is derived from PI, or its derivatives, this requirement is largely fulfilled. This is because the inositol phospholipids contain a high proportion of the C_{20} unsaturated fatty acid, arachidonic acid, at position 2 of the glycerol backbone, and, therefore, the resultant DG also contains this fatty acid. This may also be important for another reason, in that some cells (e.g. neutrophils) contain an active diacylglycerol lipase which cleaves arachidonic acid from DG, thereby releasing it for subsequent metabolism to other messenger molecules (see Chapter 7).

Synergism between 1, 2-diacylglycerol and inositol-1, 4, 5-trisphosphate

While the two arms of the inositol lipid messenger system can function independently to regulate cell activation, it is evident that they can also act synergistically under some circumstances to induce heightened cellular responses. This is best illustrated in blood platelets where stimulation of the secretion of serotonin is induced by agents such as thrombin or collagen. These agents activate PIP_2 hydrolysis by binding to specific receptors on the platelet surface and cause both Ca^{2+} mobilization and PKC activation. If, however, platelet Ca^{2+} is mobilized using, for example, the Ca^{2+} ionophore A23187, no serotonin secretion results. It is clear that Ca^{2+}-dependent reactions have been activated under these conditions, however, since a platelet protein of molecular weight 20 K (possibly the light chain of myosin) becomes phosphorylated in a calmodulin-dependent manner. Similarly, if protein kinase C is activated by treatment with a synthetic diacylglycerol (such as 1-oleoyl-2-acetylglycerol) an unidentified protein of 40-K molecular weight becomes phosphorylated but no increase in serotonin secretion results. If, however, these two stimuli are combined, a large increase in the secretion of serotonin results. Thus, there appears to be a cooperative regulation between the two separate sides of this hormone signalling system. The biochemistry of the process is still under examination and it remains uncertain how (or whether) the phosphorylation of particular platelet proteins leads to enhanced rates of secretory granule exocytosis. Similar types of synergism have also been observed in islets of Langerhans (for stimulation of insulin secretion), mast cells (histamine secretion), liver (glycogen breakdown), neutrophils (superoxide generation) and many other cell types.

Further reading

Inositol phospholipids and intracellular messenger formation – reviews

Abdel-Latif, A.A. (1986). Calcium-mobilizing receptors, polyphosphoinositides and the generation of second messengers. *Pharmacol. Rev.*, **38**, 227–72.

Altman, J. (1988). Ins and outs of cell signalling. *Nature*, **331**, 119–20.

Berridge, M.J. (1984). Inositol trisphosphate and diacylglycerol as second messengers *Biochem. J.*, **220**, 345–60.

Berridge, M.J. (1987). Inositol trisphosphate and diacylglycerol: two interacting second messengers. *Ann. Rev. Biochem.*, **56**, 159–93.

Berridge, M.J. (1987). Inositol lipids and cell proliferation. *Biochem. Biophys. Acta*, **907**, 33–45.

Berridge, M.J. and Irvine, R.F. (1984). Inositol trisphosphate, a novel second messenger in cellular signal transduction. *Nature*, **312**, 315–21.

Drummond, A.H. (1987). Lithium and inositol lipid-linked signalling mechanisms. *Trends Pharmacol. Sci.*, **8**, 129–33.

Hawthorne, J.N. (1988). Phosphoinositides and metabolic control: how many messengers? *Biochem. Soc. Trans.*, **16**, 657–66.

Hirasawa, K. and Nishizuka, Y. (1985). Phosphatidylinositol turnover in receptor mechanisms and signal transduction. *Ann. Rev. Pharmacol. Toxicol.*, **25**, 147–70.

Linden, J. and Delahunty, T.M. (1989). Receptors that inhibit phosphoinositide breakdown. *Trends Biochem. Sci.*, **10**, 114–20.

Majerus, P. (1988). Inositol phosphates: synthesis and degradation. *J. Biol. Chem.*, **263**, 3051–4.

Majerus, P.W., Connolly, T.M. and Deckmyn, H. (1986). The metabolism of phosphoinositide-derived messenger molecules. *Science*, **234**, 1519–25.

Majerus, P.W., Neufield, E.J. and Wilson, D.B. (1984). Production of phosphoinositide-derived messengers. *Cell*, **37**, 701–3.

Majerus, P.W., Wilson, D.B., Connolly, T.M. *et al.* (1985). Phosphoinositide turnover provides a link in stimulus–response coupling. *Trends Biochem. Sci.*, **10**, 168–70.

Sekar, M.C. and Hokin, L.E. (1986). The role of phosphoinositides in signal transduction. *J. Membrane Biol.* **89**, 193–210.

Taylor, C.W. (1988). Time and space – novel aspects of hormone action. *Trends Pharmacol. Sci.*, **9**, 43–5.

Williamson, J.R., Cooper, R.H., Joseph, S.K. *et al.* (1985). Inositol trisphosphate and diacylglycerol as intracellular second messengers in liver. *Am. J. Physiol.*, **428**, C203–C216.

Inositol-containing phospholipids

Berridge, M.J. (1983). Rapid accumulation of inositol trisphosphate reveals that agonists hydrolyse polyphosphoinositides instead of phosphatidylinositol. *Biochem. J.*, **212**, 849–58.

Berridge, M.J., Downes, C.P. and Henley, M.R. (1982). Lithium amplifies agonist-dependent phosphatidylinositol responses in brain and salivary glands. *Biochem. J.*, **206**, 587–95.

Bocckino, S.O., Blackmore, P.F. and Exton, J.H. (1985). Stimulation of 1, 2-diacylglycerol accumulation in hepatocytes by vasopressin, epinephrine and angiotensin II. *J. Biol. Chem.*, **260**, 14201–7.

Downes, C.P. and Wusteman, M.N. (1983). Breakdown of polyphosphoinositides and not phosphatidylinositol accounts for muscarinic agonist-stimulated inositol phospholipid metabolism in rat parotid glands. *Biochem. J.*, **216**, 633–40.

Farese, R.V., Orchard, J.L., Larson, R.E. *et al.* (1985). Phosphatidylinositol hydrolysis and phosphatidylinositol-4, 5-diphosphate hydrolysis are separable responses during secretagogue action in the rat pancreas. *Biochim. Biophys. Acta*, **846**, 296–304.

Fisher, S.K., Van Rooijen, L.A.A. and Agranoff, B.W. (1984). Renewed interest in the polyphosphoinositides. *Trends Biochem. Sci.*, **9**, 53–6.

Hawthorne, J.N. (1983). Polyphosphoinositide metabolism in excitable membranes. *Biosci. Rep.*, **3**, 887–904.

Imai, A. and Gershengorn, M.C. (1986). Phosphatidylinositol-4, 5-bisphosphate turnover is transient while phosphatidylinositol turnover is persistent in thyrotropin-releasing hormone-stimulated rat pituitary cells. *Proc. Natl. Acad. Sci. USA*, **83**, 8540–4.

Marx, J.L. (1984). A new view of rerceptor action. *Science*, **224**, 271–4.

Nishizuka, Y. (1984). Turnover of inositol phospholipids and signal transduction. *Science*, **225**, 1365–70.

Prpic, V., Blackmore, P.F. and Exton, J.H. (1982). *Myo* inositol uptake and metabolism in isolated rat liver cells. *J. Biol. Chem.*, **257**, 11315–22.

Rhodes, D., Prpic, V., Exton, J.H. and Blackmore, P.F. (1983). Stimulation of phosphatidylinositol-4, 5-bisphosphate hydrolysis in hepatocytes by vasopressin. *J. Biol. Chem.*, **258**, 2770–3.

Seyfred, M.A. and Wells, W.W. (1984). Subcellular incorporation of ^{32}P into phosphoinositides and other phospholipids in isolated hepatocytes. *J. Biol. Chem.*, **259**, 7659–65.

Weiss, S.J., McKinney, J.S. and Putney, J.W. (1982). Receptor mediated net breakdown of phosphatidylinositol-4, 5-bisphosphate in parotid acinar cells. *Biochem. J.*, **206**, 555–60.

Phosphoinositide-specific phospholipase C

Cockcroft, S., Baldwin, J.M. and Allan, D. (1984). The Ca^{2+}-activated polyphosphoinositide phosphodiesterase of human and rabbit neutrophil membranes. *Biochem. J.*, **221**, 477–82.

Irvine, R.F., Letcher, A.J. and Dawson, R.M.C. (1984). Phosphatidylinositol-4, 5-bisphosphate phosphodiesterase and phosphomonoesterase activities of rat brain. *Biochem. J.*, **218**, 177–85.

Joseph, S.K. and Williams, R.J. (1985). Subcellular localisation and some properties of the enzymes hydrolysing polyphosphoinositides in rat liver. *FEBS Lett.*, **180**, 150–4.

Rebecchi, M.J. and Rosen, O.M. (1987). Purification of a phosphoinositide-specific phospholipase C from bovine brain. *J. Biol. Chem.*, **262**, 12526–32.

Renard, D., Poggioli, J., Berthon, B. *et al.* (1987). How far does phospholipase C activity depend on the cell calcium concentration? *Biochem. J.*, **243**, 391–8.

Ryu, S.H., Cho, K.S., Lee, K.Y. *et al.* (1987). Purification and characterisation of two immunologically distinct phosphoinositide-specific phospholipase C from bovine brain. *J. Biol. Chem.*, **262**, 12511–18.

Ryu, S.H., Suh, P.G., Cho, K.S. *et al.* (1987). Bovine brain cytosol contains three immunologically distinct forms of inositol phospholipid-specific phospholipase C. *Proc. Natl. Acad. Sci. USA*, **84**, 6649–53.

Wilson, D.B., Bross, T.E., Hoffman, S.L. *et al.* (1984). Hydrolysis of polyphosphoinositides by purified sheep seminal vesicle phospholipase C enzymes. *J. Biol. Chem.*, **259**, 11718–24.

Regulation of phospholipase C by a G-protein

Baldassare, J.J. and Fisher, G.J. (1986). Regulation of membrane associated and cytosolic phospholipase C activities in human platelets by guanosine trisphosphate. *J. Biol. Chem.*, **261**, 11942–4.

Cockcroft, S. and Gomperts, B.D. (1985). Role of guanine nucleotide binding protein in the activation of polyphosphoinositide phosphodiesterase. *Nature*, **314**, 534–6.

Cockcroft, S. and Taylor, J.A. (1987). Fluoraluminates mimic guanosine $5'[\gamma$-thio] trisphosphate in activating the polyphospoinositide phosphodiesterase of hepatocyte membranes. *Biochem. J.*, **241**, 409–14.

Deckmyn, H., Tu, S.M. and Majerus, P.W. (1986). Guanine nucleotides stimulate soluble phosphoinositide-specific phospholipase C in the absence of membranes. *J. Biol. Chem.*, **261**, 16553–8.

Litosch, I. (1987). Guanine nucleotide and NaF stimulation of phospholipase C activity in rat cerebral–cortical membranes. *Biochem. J.*, **244**, 35–40.

Lo, W.W.Y. and Hughes, J. (1987). A novel cholera toxin-sensitive G-protein (Gc) regulating receptor-mediated phosphoinositide signalling in human pituitary clonal cells. *FEBS Lett.*, **220**, 327–31.

Lucas, D.O., Bajjalieh, S.M., Kowalchyk, J.A. and Martin, T.F.J. (1985). Direct stimulation by thyrotropin releasing hormone (TRH) of polyphosphoinositide hydrolysis in GH_3 cell membranes by a guanine nucleotide-modulated mechanism. *Biochem. Biophys. Res. Comm.*, **132**, 721–8.

Michell, R. and Kirk, C. (1986). G-protein control of inositol phosphate hydrolysis. *Nature*, **323**, 112–13.

Taylor, S.J. and Exton, J.H. (1987). Guanine-nucleotide and hormone regulation of polyphosphoinositide phospholipase C activity of rat liver plasma membranes. *Biochem. J.*, **248**, 791–9.

Uhing, R.J., Prpic, V., Jiang, H. and Exton, J.H. (1985). Hormone stimulated polyphosphoinositide breakdown in rat liver plasma membranes. *J. Biol. Chem.*, **261**, 2140–6.

Wakelam, M.J.O., Davies, S.A., Houslay, M.D. *et al.* (1986). Normal $p21^{N-ras}$ couples bombesin and other growth factor receptors to inositol phosphate production. *Nature*, **323**, 173–6.

Wallace, M.A. and Fain, J.N. (1985). Guanosine 5′-O-thiotriphosphate stimulates phospholipase C activity in plasma membranes of rat hepatocytes. *J. Biol. Chem.*, **260**, 9527–30.

Inositol phosphates as intracellular messengers

Authi, K.S. and Crawford, N. (1985). Inositol 1, 4, 5-trisphosphate-induced release of sequestered Ca^{2+} from highly purified human platelet intracellular membranes. *Biochem. J.*, **230**, 247–53.

Burgess, G.M., Godfrey, P.P., McKinney, J.S. *et al.* (1984). The second messenger linking receptor activation to internal Ca release in liver. *Nature*, **309**, 63–6.

Charest, R., Prpic, V., Exton, J.H. and Blackmore, P.F. (1985). Stimulation of inositol trisphosphate formation in hepatocytes by vasopressin, adrenaline and angiotensin II and its relationship to changes in cytosolic free Ca^{2+}. *Biochem. J.*, **227**, 79–90.

Delfert, D.M., Hill, S., Pershadsingh, H.A. *et al.* (1986). *Myo*-inositol-1,4,5-tri-sphosphate mobilizes Ca^{2+} from isolated adipocyte endoplasmic reticulum but not from plasma membranes. *Biochem. J.*, **236**, 37–44.

Exton, J.H. (1988). Mechanisms of action of Ca-mobilizing agonists – some variations on a young theme. *FASEB J.*, **2**, 2670–6.

Guillemette, G. (1988). Characterization of inositol trisphosphate receptors and calcium mobilization in hepatic membrane fractions. *J. Biol. Chem.*, **263**, 4541–8.

Irvine, R.F., Brown, K.D. and Berridge, M.J. (1984). Specificity of inositol trisphosphate-induced calcium release from permeabilized Swiss-mouse 3T3 cells. *Biochem. J.*, **221**, 269–72.

Irvine, R.F. and Moor, R.M. (1987). Inositol-1,3,4,5-tetrakisphosphate-induced activation of sea urchin eggs requires the presence of inositol trisphosphate. *Biochem. Biophys. Res. Commun.*, **146**, 284–90.

Joseph, S.K. (1984). Inositol trisphosphate: an intracellular messenger produced by Ca^{2+} mobilizing hormones. *Trends Biochem. Sci.*, **9**, 420–1.

Joseph, S.K. and Rice, H.L. (1989). The relationship between inositol trisphosphate receptor density and calcium release in brain microsomes. *J. Pharmacol. Exp. Ther.*, **35**, 355–9.

Michell, R.H. (1986). A second messenger function for inositol tetrakisphosphate. *Nature*, **324**, 613.

Muallem, S., Schoeffield, M., Pandol, S. and Sachs, G. (1985). Inositol trisphosphate modification of ion transport in rough endoplasmic reticulum. *Proc. Natl. Acad. Sci. USA*, **82**, 4433–7.

Prentki, M., Biden, T.J., Janjic, D. *et al.* (1984). Rapid mobilization of Ca^{2+} from rat insulinoma microsomes by inositol-1, 4, 5-trisphosphate. *Nature*, **309**, 562–4.

Spat, A., Bradford, P.G., McKinney, J.S. *et al.* (1986). A saturable receptor for ^{32}P-inositol-1,4,5-trisphosphate in hepatocytes and neutrophils. *Nature*, **319**, 514–16.

Streb, H., Bayerdorffer, E., Haase, W. *et al.* (1984). Effect of inositol-1,4,5-tri-sphosphate in isolated subcellular fractions of rat pancreas. *J. Memb. Biol.*, **81**, 241–53.

Streb, H., Irvine, R.F., Berridge, M.J. and Schulz, I. (1983). Release of Ca^{2+} from a non-mitochondrial intracellular store in pancreatic acinar cells by inositol-1,4,5-trisphosphate. *Nature*, **306**, 67–8.

Taylor, C.W. (1987). Receptor regulation of calcium entry. *Trends Pharmacol. Sci.*, **8**, 79–80.

Turner, P.R., Jaffe, L.A. and Fein, A. (1986). Regulation of cortical vesicle exocytosis in sea urchin eggs by inositol-1, 4, 5-trisphosphate and GTP-binding protein. *J. Cell Biol.*, **102**, 70–6.

Whitaker, M. (1985). Polyphosphoinositide hydrolysis is associated with exocytosis in adrenal medullary cells. *FEBS Lett.*, **189**, 137–40.

Whitaker, M. and Irvine, R.F. (1984). Inositol-1,4,5-trisphosphate microinjection activates sea urchin eggs. *Nature*, **312**, 636–9.

Inositol phosphate isomers

Balla, T., Guillemette, G., Baukal, A.J. and Katt, K.J. (1987). Metabolism of inositol-1, 3, 4-trisphosphate to a new tetrakisphosphate isomer in angiotensin-stimulated adrenal glomerulosa cells. *J. Biol. Chem.*, **262**, 9952–6.

Batty, I.R., Nahorski, S.R. and Irvine, R.F. (1985). Rapid formation of inositol 1, 3, 4, 5-tetrakisphosphate following muscarinic receptor stimulation of rat cerebral cortical slices. *Biochem. J.*, **232**, 211–15.

Biden, T.J. and Wolheim, C.B. (1986). Ca^{2+} regulates the inositol tris/tetrakisphosphate pathway in intact and broken preparations of insulin-secreting RINm5F cells. *J. Biol. Chem.*, **261**, 11931–4.

Hansen, C.A., Mah, S. and Williamson, J.R. (1986). Formation and metabolism of inositol-1,3,4,5-tetrakisphoshate in liver. *J. Biol. Chem.*, **261**, 8100–3.

Hawkins, P.T., Stephens, L. and Downes, C.P. (1986). Rapid formation of inositol-1,3,4,5-tetrakisphosphate and inositol-1,3,4-trisphosphate in rat parotid glands may both result indirectly from receptor-stimulated release of inositol-1,4,5-trisphosphate from phosphatidylinositol-4,5-bisphosphate. *Biochem. J.*, **238**, 507–16.

Michell, R. (1986). Inositol phosphates – profusion and confusion. *Nature*, **319**, 176–7.

Nahorski, S.R. and Batty, I. (1986) Inositol tetrakisphosphate: recent developments in phosphoinositide metabolism and receptor function. *Trends Pharmacol. Sci.*, **7**, 83–5.

Shears, S.B., Kirk, C.J. and Michell, R.H. (1987). The pathway of *myo*-inositol-1,3,4-trisphosphate dephosphorylation in liver. *Biochem. J.*, **248**, 977–80.

Shears, S.B., Parry, J.B., Tang, E.K.Y. *et al.* (1987). Metabolism of D-*myo*-inositol-1,3,4,5-tetrakisphosphate by rat liver, including the synthesis of a novel isomer of *myo*-inositol tetrakisphosphate. *Biochem. J.*, **246**, 139–47.

Metabolism of inositol phosphates

Biden, T.J., Comte, M., Cox, J.A. and Wollheim, C.B. (1987). Calcium–calmodulin stimulates inositol-1,4,5-trisphosphate kinase activity from insulin-secreting RINm5F cells. *J. Biol. Chem.*, **262**, 9437–40.

Downes, C.P., Mussat, M.C. and Michell, R.H. (1982). The inositol trisphosphate phosphomonoesterase of the human erythrocyte membrane. *Biochem. J.*, **203**, 169–77.

Irvine, R.F., Letcher, A.J., Heslop, J.P. and Berridge, M.J. (1986). The inositol tris/tetrakisphosphate pathway – demonstration of Ins (1,4,5)P$_3$ 3-kinase activity in animal tissues. *Nature*, **320**, 631–4.

Rana, R.S., Sekar, M.C., Hokin, L.E. and MacDonald, M.J. (1986). A possible role for glucose metabolites in the regulation of inositol-1,4,5-trisphosphate 5-phosphomonoesterase activity in pancreatic islets. *J. Biol. Chem.*, **261**, 5237–40.

Seyfred, M.A., Farrell, L.E. and Wells, W.W. (1984). Characterisation of D-*myo*-inositol-1,4,5-trisphosphate phosphatase in rat liver plasma membranes. *J. Biol. Chem.*, **259**, 13204–8.

Seyfred, M.A. and Wells, W.W. (1984). Subcellular site and mechanism of vasopressin-stimulated hydrolysis of phosphoinositides in rat hepatocytes. *J. Biol. Chem.*, **259**, 7666–72.

Storey, D.J., Shears, S., Kirk, C.J. and Michell, R.H. (1984). Stepwise enzymatic dephosphorylation of inositol-1,4,5-trisphosphate to inositol in liver. *Nature*, **312**, 374–6.

Yamaguchi, K., Hirata, M. and Kuriyama, H. (1987). Calmodulin activates inositol 1, 4, 5-trisphosphate 3-kinase activity in pig aortic smooth muscle. *Biochem. J.*, **244**, 787–91.

6 Insulin and growth factors

Introduction

Insulin occupies a unique position among those hormones that control carbohydrate metabolism in animals since it is the only agent which causes a lowering of blood glucose levels. For this reason, insulin deficiency has profound physiological consequences and leads to the development of diabetes mellitus. Insulin is also unique in that it is a hormone which affects almost all tissues of the body and is responsible for inducing a wide range of both metabolic and trophic responses in cells. It is not the purpose to consider, in detail, the metabolic systems which are regulated by insulin, but Table 6.1 lists some of the enzymes whose activity is altered after treatment of intact cells with this hormone. A brief examination of this list reveals that the effects of insulin are multiple and it is not surprising then, that much effort has gone into attempts to unravel the molecular details of insulin's signal transduction mechanism. Recent results have begun to shed light on this 'black box' area but much work is still required to show how several possible candidates are co-ordinated to form a unified signalling system for this hormone.

One area of research which has proved particularly fruitful has been the biochemistry of the insulin receptor. This has yielded several biochemical surprises and has opened up a whole new field of receptor research.

Insulin receptor: general properties

Early studies using proteolytic enzymes to modify the surface composition of fat cells revealed that the actions of insulin to enhance glucose uptake and inhibit lipolysis could be blocked by incubation of the cells with trypsin. This evidence in support of a cell surface-associated protein receptor molecule was subsequently confirmed by the demonstration that the specific binding of ^{125}I-labelled insulin to intact cells was markedly inhibited if the cells were first treated with proteases. Specific binding of this ligand could also be measured in cell membrane preparations, where it has been found that the interaction of insulin with its receptor is of high affinity ($K_D \sim 1\,\mathrm{nM}$) and exhibits a number of complex features. Thus, Scatchard analysis of the binding data produces graphs which deviate from linearity, suggesting that either there are multiple binding sites for the ligand or that the binding affinity of individual receptors varies according to the insulin concentration. This latter proposition was

Table 6.1 Examples of insulin sensitive enzymes

Enzyme	Activity change
Acetyl-CoA carboxylase	Increase
Phosphorylase kinase	Decrease
Pyruvate dehydrogenase	Increase
Pyruvate kinase	Decrease
Triacylglycerol lipase	Decrease
Glycogen synthase	Increase
Hydroxymethylglutaryl-CoA reductase	Increase
cAMP phosphodiesterase	Increase
Adenylate cyclase	Decrease

supported by evidence which showed that the rate of dissociation of ^{125}I-insulin from membrane receptors could be increased by dilution of the receptor preparation with a medium containing excess unlabelled hormone. This phenomenon has been termed 'Negative Cooperativity' (see Chapter 2) and it demonstrates that the receptors do not exist in isolation in the membrane but that they interact with one another in a dynamic manner. Thus, the binding of insulin to one receptor somehow leads to the transmission of a message to neighbouring receptor molecules which then induces a change in their hormone binding characteristics such that further insulin binding is not favoured. Interestingly, this phenomenon appears to be regulated by insulin itself, since it has been possible to synthesize analogues of the hormone which will bind to the receptor but which do not induce negative cooperativity. Examination of the structural characteristics of these analogues has revealed that a specific group of amino acid residues seems to be responsible for the negative cooperativity effect. These occupy positions 21 of the A chain and 23–26 of the B-chain of insulin and are invariant in all mammalian insulins. These residues are apparently not part of the 'bio-active' site of the molecule since analogues with modifications in the cooperative region can still elicit cellular responses. However, the extreme conservation of this sequence suggests that the ability of insulin to induce negative cooperativity in its receptor forms a very important regulatory component of the receptor signalling mechanism.

Subunit composition of the insulin receptor

The insulin receptor is a multimeric protein which has a native molecular weight of between 300 and 400 K as determined by *in vitro* hydrodynamic measurements. Covalent cross-linking of iodinated insulin to the membranes of intact cells has also revealed a radiolabelled species of 300-K molecular weight. Binding of ^{125}I-insulin to this protein exhibits a specificity which suggests that it is a true receptor molecule. Since one subunit retains most of the label following covalent attachment of ^{125}I-insulin to the receptor (the α-subunit), it is assumed to contain the hormone binding site. The molecule is a glycoprotein and can be labelled *in situ* by incubation of cells with a range of radioactive sugars, including ^{3}H-mannose and ^{3}H-fucose. This procedure also labels a second protein of molecular weight ~ 95 K which can also be immunoprecipitated by antibodies raised against the insulin receptor. This smaller component is not readily visualized by insulin-affinity labelling techniques, suggesting that it does not contain a binding site for insulin, but it is believed to be part of the receptor complex. This component is now referred to as the β-subunit. The stoichiometry of the $\alpha:\beta$ subunit ratio suggests that each receptor contains equal numbers of both type of subunit, and it is now believed that both polypeptide chains are synthesized in the cell as part of a single large precursor molecule. This was first demonstrated in 'pulse-chase' experiments where cells were briefly incubated in a medium containing ^{35}S-methionine, and the appearance of the label monitored with time by autoradiography of the separated cellular proteins. Using this technique it was found that the first identifiable protein which contained the receptor sequences is a glycoprotein which has a molecular weight of ~ 210 K. This is thought to represent a single-chain precursor which is subsequently cleaved by protease action to yield the $\alpha + \beta$ subunits. These appear to be linked together in the plasma membrane by disulphide bonds which are probably formed during folding of the precursor molecule before proteolytic cleavage takes place. Since the native receptor has a molecular weight of ~ 350 K and contains equal amounts of α and β, this suggests that the receptor is a tetramer composed of two molecules of each type of subunit, i.e. $\alpha_2:\beta_2$. This is now the generally accepted model (Fig. 6.1).

Studies with reducing agents such as dithiothreitol (DTT) have suggested that the overall receptor structure is held together by disulphide bonds. These have been classified as either 'class I' or 'class II' according to their ease of reduction and a general model for the receptor structure is shown in Fig. 6.1. In this model, each α/β dimer is held together by a relatively DTT-resistant S–S bond while the pair of dimers are linked by a second, more easily reduced, S–S bond.

In a major feat of genetic engineering, the gene for the entire insulin receptor precursor has been cloned and sequenced and its amino acid sequence deduced. This has confirmed that both subunits are encoded by a single gene and are contained within a precursor protein. The study has revealed that the α-subunit occupies the amino-terminal portion of the protein and is separated from the β-subunit by a sequence of 4 basic amino acids (Arg–Lys–Arg–Arg) which probably represent the site of proteolysis. The whole protein is 1370 amino acids in length, including an amino-terminal signal sequence.

Analysis of the amino acid sequences of the two subunits of the insulin receptor have revealed some important structural characteristics. Firstly, the α-subunit is composed of 719 largely hydrophilic amino acid residues and does not contain a membrane-spanning hydrophobic sequence. In contrast, the 620 amino acid β-subunit contains a stretch of 23 hydrophobic amino acids which probably represent a transmembrane sequence. This is located nearest to the

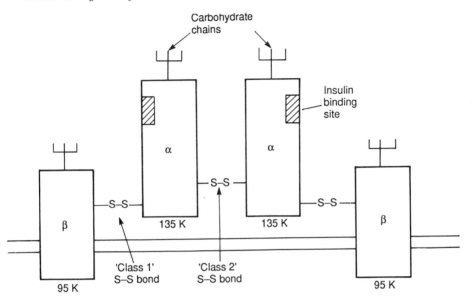

Fig. 6.1 Diagrammatic representation of the subunit composition of the insulin receptor

amino-terminus of the molecule, suggesting that about 65% of the β-subunit sequence resides on the cytoplasmic side of the membrane. The α-subunit is likely to be entirely extracellular. Both subunits possess potential glycosylation sites, although it is not known which of these are actually glycosylated in the cell.

The insulin receptor as a protein kinase

One particularly interesting feature revealed by the sequence analysis of the insulin receptor is that the β-subunit contains an intracellular domain which represents a binding site for ATP. This is important since it has been recognized that several other growth factor receptors also have homologous regions in their structures, and that these receptors can become phosphorylated upon ligand binding. This was first demonstrated for the epidermal growth factor (EGF) receptor by Dr Stanley Cohen and colleagues who also showed that the phosphorylation occurs at tyrosine residues. This is particularly interesting since such phosphorylations are very rare in cells, as the majority of cellular protein kinases show preference for either serine or threonine residues.

Treatment of liver and fat cells with insulin has been found to promote the phosphorylation of insulin receptors. Examination of the labelled proteins by PAGE under reducing conditions revealed that the phosphate becomes incorporated into a protein of 95-K molecular weight which has been identified as the β-subunit of the insulin receptor. In intact cells labelled with

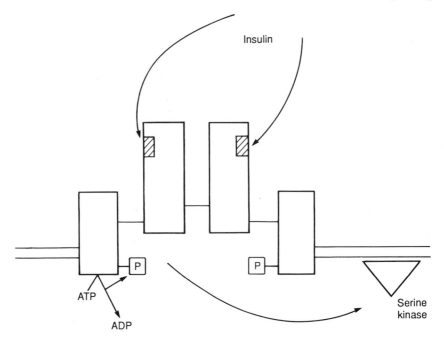

Fig. 6.2 Activation of receptor tyrosine kinase and an associated serine kinase by insulin

^{32}P, analysis of the phosphoamino acids present in the β-subunit showed the presence of phosphoserine, threonine and tyrosine under basal conditions, and an increase in the extent of phosphorylation of both serine and tyrosine following insulin administration. However, the kinetics of this increase differed between the two amino acids with the most rapid increase seen in the tyrosine residues. This is consistent with the possibility that, like the EGF receptor, the insulin receptor may also be a hormone-sensitive protein kinase with affinity for tyrosine residues.

This proposal has been confirmed by experiments utilizing highly-purified insulin receptors. These studies have shown that the receptors can catalyse the tyrosine auto-phosphorylation of their β-subunits in an insulin-sensitive manner. Serine phosphorylation is not observed in these preparations. This suggests that the insulin receptor contains an intrinsic tyrosine protein activity and that it may also be associated with a separate serine kinase in the cell (Fig. 6.2).

Biochemical characteristics of the insulin receptor protein kinase

The most studied, and apparently the best, substrate for the insulin receptor kinase is the β-subunit of the receptor itself, although in cell-free systems exogenous proteins such as histones and casein will also serve as substrates for the enzyme. ATP acts as a phosphoryl donor in the reaction mechanism and other nucleoside triphosphates cannot substitute for ATP. In the absence of insulin the receptor kinase activity is very low, but addition of the hormone leads to a very rapid increase in activity. This is brought about by a large increase in the V_{max} for ATP (by as much as 20-fold) while there is little effect on the K_m, which remains at approximately $100\mu M$ for ATP. Interestingly, the enzyme also appears to possess a second regulatory site which binds divalent metal ions. This site may be involved in lowering the K_m for ATP, since Mn^{2+} or Mg^{2+} ions can augment the extent of stimulation induced by insulin. Mn^{2+} is much more effective than Mg^{2+} in this regard while Ca^{2+} is without effect, but the physiological significance of this control by metal ions is not clear. Indeed, since under normal conditions the intracellular ATP concentration is well above the K_m value it is unlikely that the prevailing metal ion concentration will play any significant role in regulating the kinase activity.

Analysis of the phosphorylated amino acid residues in the receptor β-subunit or in any of the several exogenous protein substrates has revealed that insulin-stimulated phosphorylation occurs exclusively on tyrosine residues. Indeed, a synthetic dipeptide consisting of only Tyr and Arg can serve as a substrate for the enzyme if it is present at a sufficiently high concentration. Thus, the receptor is a true tyrosine-specific protein kinase. Kinetic studies suggest that receptor autophosphorylation is the favoured reaction for the insulin-activated kinase, and it is calculated that up to 13 molecules of phosphate can be incorporated per insulin binding site in *in vitro* assay systems. One surprising feature of the receptor autophosphorylation activity is that it is not inhibited if the receptors are diluted in order to reduce the probability of interactions between individual receptor molecules. This suggests that the reaction is truly autocatalytic and that each receptor β-subunit can phosphorylate itself.

Regulation of the receptor kinase activity

Evidence has accumulated which points to a direct regulatory role for phosphorylation of the insulin receptor in control of the kinase activity. Thus, pre-incubation of the receptor with ATP leads to a much greater kinase activity upon subsequent addition of exogenous substrate than is observed if the system is not pre-treated with ATP. This effect can be reversed by phosphatase treatment, suggesting that it reflects receptor phosphorylation. One effect of such autophosphorylation is that it converts the receptor into an insulin-insensitive state. Therefore, once phosphorylated on specific tyrosine residues the receptor becomes refractory to further insulin stimulation and is left in a fully activated state. This may be a component of the amplification mechanism of the signal transduction system, since activation of each receptor by a single binding event could lead, in principle, to phosphorylation of many molecules of substrate before deactivation occurs by the action of an appropriate phosphatase.

In intact cells, insulin receptors can also become phosphorylated on serine or threonine residues. This is not due to the kinase activity of the receptor itself but shows that the receptor can act as a substrate for other cellular kinases. Interestingly, these phosphorylation reactions may exert a negative control over the functional activity of the receptor, since they inhibit insulin-stimulated tyrosine kinase activity. This effect has been demonstrated after activation of either protein kinase A or protein kinase C in intact cells and may reflect one means by which cells coordinate incoming hormonal signals to ensure that an appropriate physiological response results.

In addition to altering the tyrosine kinase activity of the insulin receptor, these phosphorylation reactions may also play a role in the regulation of binding affinity. Activation of protein kinase C or an elevation of cAMP can both lead to a reduction in ^{125}I-insulin binding to liver and fat cells. This effect does not correlate with any loss of receptors from the cell surface and is not, therefore, a reflection of increased internalization, but seems to be due to a true decrease in the affinity of the receptors for insulin. Insulin itself can also induce a similar desensitization phenomenon in some cells, and has been found to cause an increased incorporation of phosphate into certain serine and threonine residues in its receptor. It is tempting to equate these two phenomena and to suggest that specific target amino acids can directly control receptor affinity. However, at present it is not possible to prove that a causal relationship exists, even though the evidence is highly suggestive.

Receptor tyrosine kinase activity and insulin action

In the foregoing discussion, the implicit assumption has been made that at least part of the signalling mechanism used by the insulin receptor relates to the insulin-stimulated tyrosine kinase activity of the receptor. This is not an unreasonable conclusion but there is still controversy surrounding the precise role of tyrosine phosphorylation in insulin action.

Role of insulin receptor phosphorylation

Several different lines of experimental evidence have been marshalled which support a direct role for receptor autophosphorylation in mediating some of the actions of insulin. Perhaps the most significant of these is that activation of the receptor kinase occurs over the physiological range of insulin concentrations (in the nanomolar range). This is clearly fundamental to any hypothesis linking increases in phosphorylation with other cellular events. There is also a good correlation between the ability of some modified analogues of insulin to induce receptor phosphorylation and to mimic the physiological actions of insulin. Thus, the receptor binding affinities of molecules, such as [des-alanine–des-asparagine]insulin and [des-octapeptide] insulin are similar to their respective potencies as activators of the receptor kinase and of their biological responses. Furthermore, certain antibodies can interact with the insulin receptor in such a way that they actually mimic the actions of insulin and many of these also cause an increase in receptor tyrosine kinase activity. Recent attempts have been made to substantiate the conclusions drawn from this somewhat indirect evidence by directly manipulating the receptor kinase activity in intact cells and observing the effect that this has on insulin action. One way that this has been achieved is by the introduction into cells of monoclonal antibodies raised against the insulin-receptor β-subunits. Some such antibodies specifically inhibit the receptor tyrosine kinase activity and, upon introduction into cells (by micro-injection or osmotic lysis techniques), also prevent the ability of insulin to induce a range of responses including oocyte maturation, glucose uptake, ribosomal protein phosphorylation and activation of glycogen synthesis, according to the particular cell type. In another study, the technique of site-directed mutagenesis has been employed to exchange a tyrosine residue for a non-substrate phenylalanine in the β-subunit of the insulin receptor. The 'engineered' receptor shows a marked reduction in kinase activity and when expressed in cultured cells, is also defective in the stimulation of glucose uptake. Such results therefore, support the idea that tyrosine phosphorylation plays a major role in mediating the actions of insulin. This is also given additional weight by the finding that in certain conditions that are characterized by insulin resistance *in vivo*, insulin-sensitive tissues also show a reduction in receptor tyrosine kinase activity.

However, not all of the available data entirely support the overall hypothesis. For example, some specific anti-insulin receptor antibodies have been reported to bind to the receptor in adipocytes, and to cause an increase in glucose uptake. However, no change in the extent of receptor phosphorylation could be detected in these experiments. Thus, it is not yet proven that the actions of insulin absolutely require tyrosine kinase activation and it may be that insulin receptors use several signal transduction mechanisms, of which the tyrosine kinase activity is only one.

Cellular substrates for the insulin-receptor kinase

The fact that the insulin receptor serves as a very good substrate for its own intrinsic kinase, in a reaction which may well be intramolecular, has already been discussed. It is possible then, that in the cell, this reaction is the most important for signal transduction, and that when phosphorylated the receptor becomes capable of interacting with some other potential messenger system(s) in a manner which the 'native' molecule cannot. However, an alternative possibility is that the receptor kinase, in addition to phosphorylating itself, also acts upon other specific cellular proteins to bring about some direct alteration of their individual properties. Indeed, the receptor can phosphorylate some exogenous proteins *in vitro* but these are certainly not substrates in intact cells. This has prompted a search for other putative substrates, and a number of positive findings have now been reported. Liver and brown adipose tissue both contain a glycoprotein of about 110-K molecular weight which appears to be a substrate for the insulin receptor kinase. Phosphorylation of this protein in response to insulin can only be detected after prior receptor phosphorylation, and the reaction exhibits appropriate dose–response characteristics with respect to insulin concentration. Intriguingly, the protein can also be phosphorylated by the activated epidermal growth factor (EGF) receptor (which is also a tyrosine kinase; see below) which may point to a role in cell growth regulation. However, the precise functional role of this substrate has not yet been determined. A variety of other substrate proteins has now been found in insulin sensitive tissues, including human placenta, rat adipocytes and hepatoma cells. These range in size from 46 K to 185 K but, to date, no functional activity has been positively assigned to any of them. It has been speculated that one of the proteins may be a member of the lipocortin family, which are involved in the regulation of phospholipase A_2 activity (see Chapter 7). However, confirmation of this idea and elucidation of its significance, if true, must await the results of further experiments.

Activation of a serine kinase by insulin

In intact cells it has been found that insulin promotes an increase in phosphate incorporation into both tyrosine and serine residues of its receptor. Tyrosine phosphorylation is the earlier event, as might be anticipated for an autocatalytic reaction, but significant serine phosphorylation can also be observed at later times. The serine kinase activity can be measured in crude receptor preparations but is lost upon more rigorous purification, indicating that it is probably associated with, but is not integral to, the receptor. Two schemes have been proposed to explain the relationship between the two kinase activities in insulin action. One of these envisages that each of the enzymes controls a separate component of insulin action with, for example, the tyrosine kinase exerting its influence on growth control and the serine kinase acting independently to regulate metabolic events. The second proposal suggests a sequential activation process in which both enzymes cooperate to bring about the several different physiological responses induced by insulin. Further characterization of the receptor-associated serine kinase will be required in order to help distinguish between these possibilities, and to show how its activity is regulated by insulin. Could it, perhaps, be one of the substrates for the tyrosine kinase?

Control of cAMP metabolism by insulin

In addition to activating its own signal transduction pathway, insulin can also impinge on other messenger systems. Thus, it has been recognized for quite a long time that treatment of liver cells with insulin leads to inhibition of agonist-induced cAMP accumulation. This mechanism may contribute to the ability of insulin to switch liver into an anabolic state in which nutrient storage is favoured. It appears that two distinct mechanisms may account for these effects of insulin, involving actions at the level of both cAMP synthesis and degradation.

Regulation of adenylate cyclase by insulin

Insulin appears to exert a direct inhibitory control over the activity of adenylate cyclase in liver cells. The magnitude of this effect is rather modest, representing only about 30% inhibition in isolated plasma membrane preparations, but this could still contribute to the lowering of cAMP levels in intact cells. This effect of insulin can be completely overcome by treatment of the cells with the ADP ribosyltransferase toxin of *Bordetella pertussis*. This agent catalyses the transfer of ADP-ribose from NAD^+ to the α-subunit of a guanine-nucleotide-binding protein (G_i) which mediates hormonal inhibition of adenylate cyclase (see Chapter 3 for a more complete discussion of G_i). The effect of the toxin is to prevent the dissociation of G_i into its constituent subunits, a reaction which is necessary in order for cyclase inhibition to be manifest. Since pertussis toxin also alleviates the effects of insulin on adenylate cyclase, this could be taken as evidence that insulin can activate G_i. This would then suggest that one mechanism by which insulin controls intracellular events is via a guanine-nucleotide regulatory protein, possibly G_i, and that the insulin receptor must somehow interface with the G-protein in the plasma membrane. Surprisingly, however, the inhibitory effect of insulin on adenylate cyclase activity can also be prevented by treatment of cells with a second bacterial toxin having ADP-ribosyltransferase activity – cholera toxin (Chapter 3). This toxin does not use G_i as a substrate and these results may indicate that the insulin receptor interacts with a separate, novel G-protein which is sensitive to both toxins to bring about its effects on adenylate cyclase. This idea is supported by recent evidence that the expression of the gene for G_i is, itself, subject to regulation by insulin. Thus, in diabetic animals the levels of G_i in liver are greatly decreased by comparison with normal animals, an effect which is reversed upon insulin treatment. Despite the reduction in G_i, insulin can still inhibit adenylate cyclase in cells from diabetic animals which is consistent with the intervention of another G-protein. This molecule remains to be positively identified but it has tentatively been assigned the name G_{ins} (Fig. 6.3).

One possibility is that G_{ins} may be a member of the *ras* protein family which are guanine-nucleotide-binding proteins whose precise function(s) in the cell remains a mystery. The *ras* gene codes for a protein of 21-K molecular weight (p21) and highly specific monoclonal antibodies have now been raised against a synthetic peptide derived from this molecule. These antibodies completely prevent the ability of insulin to induce maturation of *Xenopus* oocytes when micro-injected into the cells, which is highly suggestive of a role for the 21 K protein in mediating the effects of insulin in these cells. In addition, it has been found that in liver, cholera toxin can ADP-ribosylate a protein of molecular weight 25 K in a manner which is sensitive to insulin. This protein might represent G_{ins} and is sufficiently close in size to suggest that it could be p21 or a closely related molecule.

All signal transducing G-proteins which have so far been characterized in detail are multi-subunit complexes which dissociate upon interaction with an agonist-occupied receptor in a GTP-dependent manner. Whether this is also true of the putative G_{ins} is not known. However, p21 is reported to be a

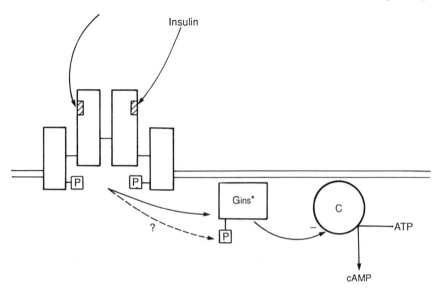

Fig. 6.3 Possible role for a G-protein in the regulation of adenylate cyclase by insulin

substrate for the insulin receptor tyrosine kinase *in vitro*, which raises the possibility that the insulin receptor could control the activation of this G-protein by a phosphorylation mechanism. If this proves to be the case then it will be an example of one signal transduction system (kinase activation) which is utilized by a hormone to control at least one (perhaps more?) other separate signalling mechanism. This would be a convenient and economical means to regulate several independent cellular functions via a single receptor.

Regulation of cAMP-degradation by insulin

cAMP breakdown is controlled by phosphodiesterase enzymes which hydrolyse the phosphodiester bond at position 3 of the ribose ring of 3':5'-cAMP to yield 5'-AMP (see Chapter 4). Many cells contain multiple forms of this enzyme and liver contains at least four distinct types. These can be fractionated on density gradients, and using this technique, it has been shown that insulin causes a stable increase in the activity of two separate forms of the enzyme. One of these is located in the hepatocyte plasma membrane and the second is associated with an intracellular membrane fraction. The plasma-membrane enzyme may be subject to regulation by an insulin-sensitive G-protein since it can also be activated by cholera toxin by a mechanism which is not related to any direct change in cell cAMP levels. It would be tempting to equate this protein with the hypothetical G_{ins}, but in this context it may be significant that pertussis toxin does not modify the ability of insulin to activate the plasma-membrane enzyme. Since the existence of G_{ins} was postulated on the basis of its susceptibility to modulation by pertussis toxin, this casts doubt

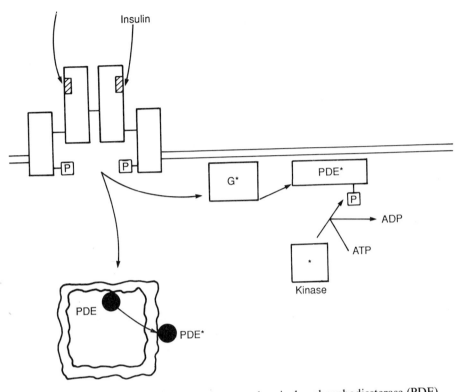

Fig. 6.4 Regulation of plasma membrane and vesicular phosphodiesterase (PDE) enzymes by insulin

on the possibility that G_{ins} could also be responsible for mediating the effects of insulin on the phosphodiesterase. Isolation of the plasma-membrane phosphodiesterase after activation by insulin reveals that part of the activation process involves a change in the phosphorylation state of the enzyme. This is seen as an increase in serine phosphorylation and is not, therefore, mediated directly by the insulin receptor itself. It may reflect the action of a cellular protein kinase which is only able to phosphorylate the enzyme after some change in its properties has been induced by the binding of insulin to its receptor (perhaps a conformational change mediated by a G-protein). Further work is still needed to clarify the present confusion surrounding the mechanism of activation of the plasma membrane phosphodiesterase by insulin (Fig. 6.4). Unfortunately, it is even less clear how insulin brings about activation of the intracellular form of phosphodiesterase. The activation process can be selectively inhibited by agents which disrupt lysosomal function, a manipulation which is without effect on the activation of the plasma membrane enzyme. This suggests that insulin controls each enzyme by an independent mechanism. One possibility is that this vesicular enzyme undergoes a translocation in response to insulin, whereby the enzyme migrates from an unidentified site of sequestration to a specific (but also unidentified) membrane compartment in which it becomes activated. The molecular details of this process also remain obscure, but, once again, a pertussis toxin sensitive G-protein has been implicated, possibly in regulation of the translocation step.

Second messengers for the actions of insulin

One early event which is central to the regulation of cellular metabolism by insulin is activation of the enzyme pyruvate dehydrogenase (PDH). This occurs within minutes in insulin-treated cells and cannot be attributed to hormone internalization which occurs with a rather slower time course. Since PDH is an exclusively mitochondrial enzyme, it is necessary to postulate that its activation by insulin must occur by the passage of some 'message' from the plasma membrane to the intracellular organelle. In contrast to the activation of the plasma-membrane phosphodiesterase by insulin, which is associated with an increase in enzyme phosphorylation (see above), activation of PDH correlates with a loss of phosphate from three specific sites in the α-subunit of the enzyme. This has been attributed to the insulin-induced activation of a PDH–phosphatase enzyme. Activation of PDH by insulin is stable and persists during preparation of subcellular fractions from hormone-stimulated cells. For this reason, one might imagine that it would be relatively easy to study the changes caused by insulin in isolated mitochondria and to identify the mechanisms involved. This, however, has proved to be extremely difficult, and the issue remains largely unresolved. The intrinsic activity of PDH phosphatase is not directly altered in insulin-treated tissue but its activity can be modulated by changes in several potential messengers, including Ca^{2+} and Mg^{2+} ions and the mitochondrial redox potential. However, careful experimentation has gradually excluded each of these from the list of possible candidates for the insulin second messenger. This has prompted the search for another 'second messenger' of insulin action which might be responsible for causing an increase in PDH phosphatase activity. Larner and colleagues claim to have identified such an activity in extracts of skeletal muscle and they have attributed this to a small peptide with a molecular weight of about 1200. Similar results have also been reported by Jarrett and co-workers who have obtained their messenger by treatment of muscle and fat cell membranes with insulin. These fractions are also reported to cause phosphodiesterase activation and adenylate cyclase inhibition under appropriate conditions, suggesting that the need to invoke G-protein mechanisms (see above) may be over-complex. Unfortunately, the situation is complicated by the apparent existence of multiple forms of the putative messenger, some of which can act antagonistically to one-another in certain assay systems (Table 6.2). Moreover, although the activity of these agents was first described in the mid-1970s, it has not yet proved possible to definitively characterize in chemical terms the molecule(s) which possess the messenger activity.

Table 6.2 Enzymes which are affected by the putative insulin second messenger

Enzyme	Effect on activity
Protein kinase A	Decrease
Adenylate cyclase	Decrease
cAMP phosphodiesterase	Increase
Pyruvate dehydrogenase	Increase
Glycogen synthase	Increase
Glucose-6-phosphatase	Decrease

Chemical characteristics of insulin second messengers

Despite the variations in activity described for the different insulin messenger preparations there is some consensus that at least one of the active components has a molecular weight of less than 2000, is stable to acid pH and to heating, and is sensitive to treatment with proteases. In addition, the ability of this compound to mimic the actions of insulin in broken tissue preparations is sensitive to neuraminidase, suggesting that the molecule may also contain carbohydrate. Some evidence has also implicated a lipid component in the structure of the messenger.

Recently, interest has once again been aroused in this field by the demonstration that insulin promotes the rapid hydrolysis of a specific membrane glycolipid in cultured muscle cells. This molecule can be radiolabelled by incubation of cells with either ^{3}H-inositol or ^{3}H-glucosamine and, in pre-labelled cells, addition of insulin is followed by a rapid loss of radioactivity from the membrane and the appearance of water-soluble products. Diacylglycerol is also generated during this process. The aqueous products are reported to be capable of inducing inhibition of adenylate cyclase and activation of cAMP-phosphodiesterase and PDH. They have therefore been equated with at least one type of the putative insulin second messengers. This overall scheme bears a striking resemblance to the $PIP_2/IP_3/Ca^{2+}$-mobilization messenger system and, like that system, also requires the activation of a specific phospholipase C in response to the hormone.

It has been well estabished that insulin does not activate the phospholipase C activity which controls PIP_2 hydrolysis in the plasma membrane of many cell types, and this suggests that an entirely separate enzyme must participate in the generation of the particular insulin 'mediator'. One potential candidate enzyme has been identified by Saltiel and co-workers in a plasma membrane fraction prepared from rat liver. This enzyme catalyses the hydrolysis of a glycosyl-phosphatidylinositol structure that is found in cell membranes, and yields diacylglycerol and a water-soluble inositol-phosphate-glycan (IPG) as products. The biochemical characteristics of this enzyme activity suggest that it is not the same phospholipase C as that which catalyses PIP_2 hydrolysis in stimulated cells, and it could certainly be a potential target for regulation by insulin. The endogenous substrate for this insulin-sensitive phosphilipase C may well be a glycosyl-phosphatidylinositol molecule which is involved in anchoring certain proteins to the plasma membrane of cells (Fig. 6.5). This structure contains glucosamine which would be consistent with the results of a preliminary characterization of the 'mediator' molecule, which also appears to contain this moiety within its structure. However, the 'mediator' does not appear to contain any ethanolamine or amino acids, both of which are components of the lipid-linked precursor, and would be released as part of the IPG structure under the influence of phospholipase C. Furthermore, the tentative assignment of a molecular weight of about 1000 to the active species derived from insulin treated tissue suggests that the initial product of phospholipase C hydrolysis (which is much larger than this) must be further modified in order to give rise to the IPG 'messenger' molecule. In this context, it has been observed that the insulin-induced release of IPG correlates with the liberation of two other molecules, alkaline phosphatase and heparan sulphate from cell membranes. This is significant since both of these proteins are anchored to the plasma membrane by a glycosyl-phosphatidylinositol molecule.

One major problem with this scheme, however, is that the membrane-anchoring glycolipid structure is located in the outer leaflet of the plasma membrane. Therefore, upon hydrolysis, any mediator substance would be released into the extracellular fluid. This does not seem very appropriate for a compound which is purported to be part of a transmembrane signalling mechanism! In an attempt to circumvent this difficulty a working hypothesis has been proposed in which insulin-sensitive tissues are considered to possess surface receptors which can bind the inositol and protein containing initial hydrolysis product. Hence, according to this scheme, upon release from the outer leaflet of the plasma membrane, this molecule immediately binds to a nearby receptor or transport protein and is translocated across the membrane for release into the cytoplasm. Part of this mechanism (Fig. 6.6) is also envisaged to result in the cleavage of IPG from the remaining components of the original structure. This sequence of events remains highly speculative at present and will need much careful experimental examination before it can be accepted as a viable model for insulin action.

One requirement of the IPG hyothesis is that the insulin receptor must be coupled to the specific phospholipase C which is responsible for initiating glycolipid hydrolysis. In the light of recent developments, it would be surprising if a G-protein were not implicated in the mechanics of this process, and it has, indeed, been reported that the production of IPG in response to insulin is inhibited by pre-incubation of the tissue with pertussis toxin. The identification of the toxin substrate will be awaited with interest – could it perhaps be G_{ins}?

It is evident from the above discussion that the mechanisms by which the insulin receptor propagates a signal across the cell membrane are complex.

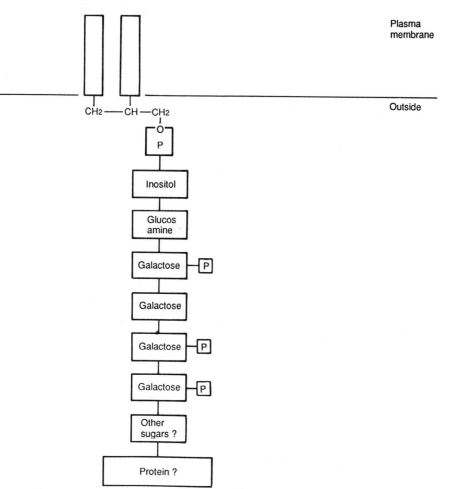

Fig. 6.5 Possible structure of an insulin sensitive glycolipid

The hormone induces a number of largely independent physiological responses and it seems probable that a variety of different signalling mechanisms are used to mediate these effects. At present it is still not possible to construct a unifying hypothesis to explain how insulin works and it seems likely that tyrosine phosphorylation, G-proteins and second messengers will all need to be incorporated. Clearly, the coupling of the insulin receptor to some or all of these processes represents an extraordinary feat of bio-engineering.

134

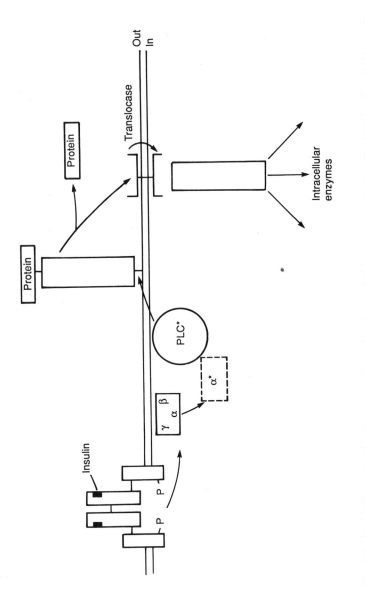

Fig. 6.6 Possible mechanism for generation of an intracellular messenger for insulin. PLC = phospholipase C.

Epidermal growth factor

Epidermal growth factor (EGF) was first identified by Stanley Cohen as an agent which can induce premature developmental changes in mice. The active principle was localized to the submaxillary gland in this species, and this rather enigmatic observation still awaits complete explanation. EGF has now been established as a potent mitogen which regulates the rate of DNA synthesis in many cell types. It is a single-chain polypeptide composed of 53 amino acids and has been most studied in the mouse. However, it does have a human equivalent which is a protein that had previously been termed urogastrone as it was originally recognized as an inhibitor of gastric acid secretion in man, but it is now believed to represent authentic human EGF.

The EGF receptor

Cell surface binding sites for EGF have been demonstrated using EGF labelled with either ^{125}I or ferritin, and such studies have also revealed that the binding of the ligand triggers the re-distribution of surface receptors into defined clusters on the cell surface. This is followed by internalization of the ligand–receptor complex and transfer of EGF to the lysosomal compartment of the cell. The demonstration that such an EGF uptake system is present in cells prompts the obvious question as to whether this forms a necessary part of the signalling mechanism by which EGF promotes DNA synthesis in cells or whether receptor internalization is a separate event. The answer to this problem has largely been provided by the isolation of a human carcinoma cell line (known as A.431) which expresses a very large number of EGF receptors on each cell. Membrane preparations from these cells can bind EGF and have been shown to possess a receptor-linked signal-transduction mechanism which can be activated on EGF binding (see below). Thus, receptor internalization is not an obligatory step in the mediation of EGF action.

The EGF receptor (Fig. 6.7) has been cloned and characterized in detail and is a glycoprotein with a molecular weight of 170 K. The receptor comprises a single polypeptide chain which traverses the plasma membrane via a single sequence of about 23 hydrophobic amino acids. The receptor therefore, has both extracellular and intracellular domains, the latter of which contains a sequence of amino acids which appears to comprise a binding site for ATP. In this respect, the EGF receptor bears obvious similarity to the insulin receptor (see above). This similarity also extends to the ability of both receptors to act as protein kinases when containing an appropriate ligand bound at an extracellular binding site. Indeed, the EGF receptor was the first to be definitively shown to contain both the receptor and kinase domains on the same molecule.

Protein kinase activity of the EGF receptor

Early work on the specificity of the EGF receptor protein kinase was interpreted as suggesting that the kinase preferentially phosphorylates

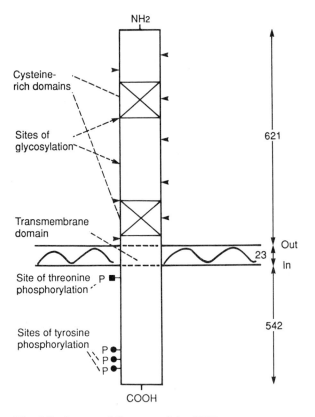

Fig. 6.7   Structural features of the EGF receptor

proteins on threonine residues. However, it soon became apparent that the electrophoretic method which had been used to determine this specificity did not resolve threonine from tyrosine. When more rigorous techniques were employed, the phosphorylation was found to be exclusive to tyrosine residues. Thus, the EGF receptor is a tyrosine-specific protein kinase. Activation of the kinase activity results from the binding of a single EGF molecule and, like the insulin receptor, is associated with an increase in the V_{max} of the reaction rather than with any change in the K_m for ATP (which is extremely low at $\sim 2.5\,\mu M$). The receptor preferentially phosphorylates itself by an intramolecular mechanism.

The molecular details of the mechanism by which ligand binding leads to kinase activation are not clear, and there is controversy regarding the precise requirements for activation. In particular, it is not certain how a binding event in one receptor domain can transmit information through a short, largely inflexible, membrane domain to induce (presumably) a conformational change in the intracellular part of the molecule which leads to enzyme activation. An alternative possibility is that the observed aggregation of receptors into particular regions of the cell surface may somehow participate in stimulation of the kinase activity by a cooperative mechanism. It might be supposed that such a mechanism would be inoperative in isolated membrane preparations which are largely devoid of the normal cytoskeletal elements present in intact cells, but biophysical evidence suggests that EGF can still induce receptor aggregation in isolated membranes. The major evidence which favours an aggregation-dependent activation mechanism relates to the observation that some anti-EGF-receptor antibodies can elicit kinase activation upon interaction with the receptor. This requires that either multivalent antibodies are present or that univalent molecules are cross-linked, suggesting that some degree of receptor association is necessary. However, whichever mechanism is eventually found to be correct, it will still be necessary to explain in molecular terms how the transmembrane signalling process occurs – at present this remains a mystery.

Substrates for the EGF receptor kinase

The EGF receptor has been identified as a phosphoprotein *in vivo*, but most of the native phosphoamino acids are either serine or threonine. Binding of EGF leads to incorporation of phosphate into three specific tyrosine residues near the carboxy-terminus of the molecule, demonstrating that the receptor itself acts as a substrate for its own intrinsic kinase. In the case of the insulin receptor, autophosphorylation leads to a direct enhancement of kinase activity but it is less clear whether this is also true for the EGF receptor.

A number of exogenous proteins will serve as substrates for the EGF receptor kinase, including gastrin, angiotensin II, growth hormone and myosin light chains but the identity of its endogenous substrates is less firmly established. These may include members of the lipocortin family (cf. the insulin receptor kinase) and an unidentified protein of 42-K molecular weight. This protein is of special interest since it seems to be a particularly good substrate for the kinase and because it is also phosphorylated by other agents which, like EGF, induce cell proliferation. Identification and characterization of this molecule may well provide clues about the regulation of cell growth and differentiation if, indeed, it is a true target for the kinase.

It is still equivocal whether EGF stimulation of receptor tyrosine kinase activity is essential to the induction of physiological responses by EGF. These include a rapid increase in the free cytosolic Ca^{2+} concentration, an increase in cytosolic pH, changes in gene transcription rates and DNA replication. They occur on very different time scales (ranging from seconds to hours) and there is evidence from receptor antibody studies that at least some of these processes may not require receptor kinase activation. However, a recent study has employed site-directed mutagenesis of the EGF receptor gene to modify the structure of the receptor such that it can no longer function as a kinase enzyme.

This involved replacement of a lysine residue in the ATP-binding site with a methionine and expression of the mutant receptor in a cell line which normally lacks EGF receptors. The receptor was correctly inserted into the plasma membrane, and cells transfected with a gene coding for the unmodified receptor responded to EGF treatment with increased rates of gene transcription and cell division. EGF also promoted a rapid increase in the concentration of cytosolic free Ca^{2+} in these cells. In contrast, cells expressing the mutant receptor failed to exhibit any of these responses when stimulated with EGF. This evidence strongly implicates the EGF receptor tyrosine kinase in the mechanism of signal transduction.

Modulation of the EGF receptor by protein kinase C (PKC)

In addition to serving as a substrate for its own receptor kinase, the EGF receptor can also be phosphorylated by protein kinase C. This response probably forms part of a wider mechanism which allows cells to coordinate their responses to extracellular signals. Protein kinase C phosphorylates a threonine residue which is close to the transmembrane domain of the receptor and induces two responses as a consequence: (1) decreased EGF binding and (2) reduced tyrosine kinase activity. The reduction in EGF binding may be mediated by a combination of increased receptor internalization and by a reduction in the affinity of the binding site for EGF. Since treatment of cells with EGF can lead to protein kinase C activation, this represents a feedback control mechanism which will tend to minimize prolonged activation of stimulated cells.

Growth-factor receptors and oncogenes

It has become clear in recent years that certain growth factor receptors may be related to specific genes that are activated by particular oncogenic viruses, and which are associated with the transforming potential of those viruses. For example, the oncogene which becomes active in cells infected with avian erythroblastosis virus is known as *erb-B* and codes for a protein which shows up to 95% homology with the cytoplasmic region of the EGF receptor. This had led to speculation that *erb-B* may function as a non-regulated tyrosine kinase in cells and that this may account for its ability to induce transformation.

A second EGF-like oncogene has also been identified and is known as *neu*. This gene codes for a large protein ($\sim$ 185-K molecular weight) which is about 50% identical to the EGF receptor sequence. The true cellular equivalent of this viral gene may be different from the EGF receptor and has not yet been functionally identified, although it is suggested that it may also be a growth-factor receptor.

Several growth-factor receptors have now been analysed in detail and they seem to comprise three basic groups. One group contains the EGF receptor and several other molecules which show homology to it, including the receptor for platelet-derived growth factor (PDGF). This group comprises molecules that are all single-chain structures which possess intrinsic tyrosine kinase activity and have marked homology to identifiable viral oncogenes. The second group contains the receptors for insulin and insulin-like growth factor-I (IGF-I) which are composed of separate subunits bearing distinct ligand-binding and tyrosine kinase domains. These may also have oncogene equivalents, including *sic* and *fins*. The third group includes the receptors for nerve growth factor (NGF) and interleukin-2 (IL-2) which differ from the previous classes in that they lack an identifiable tyrosine kinase domain. A particularly interesting recent finding is that the receptor for another insulin-like growth factor, IGF-II, which is believed to mediate at least some of the growth promoting activities of this molecule, has a structure which is unlike that of any other known growth-factor receptor. The primary amino acid sequence of this molecule has recently been deduced from the cloned gene, and it is shown to be a large protein (~ 300-K molecular weight) composed of nearly 2500 amino acids. Interestingly, only 164 of these appear to reside on the cytoplasmic face of the plasma membrane. The most remarkable thing is that the receptor bears striking homology with another receptor which plays a role in targeting of proteins bearing mannose-6-phosphate residues to lysosomes. Indeed, the homology is so great that these may be the same protein. Hence, this receptor could serve several different functions in separate regions of the cell, and could interact with multiple ligands which makes it unique as a signalling molecule. The mechanism by which it transduces the IGF-II signal is not yet understood but, on the basis of the amino acid sequence, is unlikely to be related to any direct protein kinase activity. The receptor may perhaps interact with another molecule to bring about its effect. Clearly, intrinsic tyrosine kinase activity is not all important for growth promotion in cells, and other signalling systems can also be used. Equally, however, the observation that an increasing number of oncogene products have been shown to be related to growth factor receptors with tyrosine kinase activity suggests that this is an extremely important intracellular signalling mechanism. Furthermore, the possibility that the transforming potential of oncogenes may be related to inappropriate regulation of the tyrosine kinase activity, emphasizes the fine balance between a coordinated cellular response that is induced by an agonist and the uncontrolled cell growth resulting from virus infection.

Further reading

The insulin receptor – structure and function

Czech, M.P. (1984). New perspectives on the mechanism of insulin action. *Rec. Prog. Horm. Res.*, **40**, 347–73.

Czech, M.P. (1985). The nature and regulation of the insulin receptor: structure and function. *Ann. Rev. Physiol.*, **47**, 357–81.

Gammeltoft, S. (1984). Insulin receptors: binding kinetics and structure function relationship of insulin. *Physiol. Rev.*, **64**, 1321–78.

Gammeltoft, S., Kowalski, A. and Fehlmann, M. *et al.* (1984). Insulin receptors in rat brain: insulin stimulates phosphorylation of its receptor β-subunit. *FEBS Lett.*, **172**, 87–91.

Gammeltoft, S. and Van Obberghen, E. (1986). Protein kinase activity of the insulin receptor. *Biochem. J.*, **235**, 1–11.

Gazzano, H., Kowalski, A. and Fehlmann, M. *et al.* (1983). Two different protein kinase activities are associated with the insulin receptor. *Biochem. J.*, **216**, 575–82.

Goldfine, I.D. (1987). The insulin receptor: Molecular biology and transmembrane signalling. *Endo. Rev.*, **8**, 235–55.

Kahn, C.R. (1985). The molecular mechanism of insulin action. *Ann. Rev. Med.*, **36**, 429–51.

Kin-Tak, Y. and Czech, M.P. (1984). Tyrosine phosphorylation of the insulin receptor subunit activates the receptor associated tyrosine kinase activity. *J. Biol. Chem.*, **259**, 5277–86.

Kin-Tak, Y., Khalaf, N. and Czech, M.P. (1987). Insulin stimulates a membrane-bound serine kinase that may be phosphorylated on tyrosine. *Proc. Natl. Acad. Sci. USA*, **84**, 3972–6.

Machicao, F., Haring, H., White, M.F. *et al.* (1987). An M_r 180 000 protein is an endogenous substrate for the insulin-receptor-associated tyrosine kinase in human placenta. *Biochem. J.*, **243**, 797–801.

Machicao, F., Urumow, T. and Wieland, O.H. (1983). Evidence for phosphorylation of actin by the insulin receptor-associated protein kinase from human placenta. *FEBS Lett.*, **163**, 76–80.

Nemenoff, R.A., Kwok, Y.C., Shulman, G.I. *et al.* (1984). Insulin-stimulated tyrosine protein kinase: characterisation and relation to the insulin receptor. *J. Biol. Chem.*, **259**, 5058–65.

Phillips, S.A., Perotti, N. and Taylor S.I. (1987). Rat liver membranes contain a 120 kDa glycoprotein which serves as a substrate for the tyrosine kinases of the receptors for insulin and epidermal growth factor. *FEBS Lett.*, **212**, 141–4.

Sadoul, J.L., Peyron, J.F., Ballotti, R. *et al.* (1985). Identification of a cellular 110 000-Da protein substrate for the insulin-receptor kinase. *Biochem. J.*, **227**, 887–92.

Ullrich, A., Bell, J.R., Chen, E.Y. *et al.* (1985). Human insulin receptor and its relationship to the tyrosine kinase family of oncogenes. *Nature*, **313**, 756–61.

Van Obberghen, E. (1984). The insulin receptor: its structure and function. *Biochem. Pharmacol.*, **33**, 889–96.

Role of tyrosine kinase activity in insulin action

Czech, M. (1988). Insulin receptor signalling—activation of multipile serine kinases. *J Biol. Chem.*, **263**, 11017–20.

Espinal, J. (1988). What is the role of the insulin receptor tyrosine kinase? *Trends Biochem. Sci.*, **13**, 367–9.

Forsayeth, J.R., Caro, J.F., Sinha, M.K. *et al.* (1987). Monoclonal antibodies to the human insulin receptor that activate glucose transport but not insulin receptor kinase activity. *Proc. Natl. Acad. Sci. USA*, **84**, 3448–51.

Grigorescu, F., Flier, J.S. and Kahn, C.R (1984). Defect in insulin receptor phosphory-lation in erythrocytes and fibroblasts associated with severe insulin resistance. *J. Biol. Chem.*, **259**, 15003–6.

Grunberger, G., Zick, Y. and Gordon, P. (1984). Defect in phosphorylation of insulin receptors in cells from an insulin resistant patient with normal insulin binding. *Science*, **223**, 932–4.

Karasik, A. (1988). Lipocortins I & II as substrates for the insulin receptor kinase in rat liver. *J. Biol. Chem.*, **263**, 11862–7.

Morgan, D., Ho, L., Korn, L.J. *et al.* (1986). Insulin action is blocked by a monoclonal antibody that inhibits the insulin receptor kinase. *Proc. Natl. Acad. Sci. USA*, **83**, 328–32.

Morgan, D. and Roth, R.A. (1987). Acute insulin action requires insulin receptor kinase activity. Introduction of an inhibitory monoclonal antibody into mammalian cells blocks the rapid effects of insulin. *Proc. Natl. Acad. Sci. USA*, **84**, 41–5.

Sale, G.J. (1988). Recent progress in our understanding of the mechanism of action of insulin. *Int. J. Biochem.*, **20**, 897–908.

Simpson, I.A. and Hedo, J.A. (1984). Insulin receptor phosphorylation may not be a prerequisite for acute insulin action. *Science*, **223**, 1301–4.

Strout, H.V. (1988). Protein phosphotyrosine phosphatase with activity against the insulin receptor. *Biochem. Biophys. Res. Commun.*, **151**, 633–40.

G-proteins as mediators of insulin action

Gawler, D. and Houslay, M.D. (1987). Insulin stimulates a novel GTPase activity in human platelets. *FEBS Lett*, **216**, 94–8.

Heyworth, C.M. and Houslay, M.D. (1983). Insulin exerts actions through a distinct species of guanine nucleotide regulatory protein: inhibition of adenylate cyclase. *Biochem. J.*, **214**, 547–52.

Heyworth, C.M., Whetton, A.D., Wong, S. *et al.* (1985). Insulin inhibits the cholera-toxin-catalysed ribosylation of a M_r 25 000 protein in rat liver plasma membranes. *Biochem. J.*, **228**, 593–603.

Housley, M.D. (1985). Insulin glucagon and the receptor-mediated control of cyclic AMP concentrations in liver. *Biochem. Soc. Trans.*, **14**, 183–93.

Korn, L.J., Siebel, C.W., McCormick, F. *et al.* (1987). ras p21 as a potential mediator of insulin action in *Xenopus* oocytes. *Science*, **236**, 840–3.

O'Brien, R.M., Siddle, K., Houslay, M.D. *et al.* (1987). Interaction of the human insulin receptor with the *ras* oncogene product p21. *FEBS Lett*, **217**, 253–9.

Intracellular messengers for insulin

Alemany, S., Mato, J.M. and Stralfors, P. (1987). Phospho-dephospho control by insulin is mimicked by a phospho-oligosaccharide in adipocytes. *Nature*, **330**, 77–9.

Cheng, K. and Larner, J. (1985). Intracellular mediators of insulin action. *Ann. Rev. Physiol.*, **47**, 405–24.

Denton, R.M. (1986). Early events in insulin actions. *Adv. Cyc. Nuc. Prot. Phos. Res.*, **20**, 293–342.

Espinal, J. (1987). Mechanism of insulin action. *Nature*, **328**, 574–5.

Fox, J.A., Soliz, N.H. and Saltiel, A.R. (1987). Purification of a phosphatidylinositol-glycan-specific phospholipase C from liver plasma membranes: a possible target of insulin action. *Proc. Natl. Acad. Sci. USA*, **84**, 2663–7.

Gottschalk, W.K. (1988). Insulinomimetic effects of the polar head group of an insulin sensitive glycophospholipid *Archiv. Biochem. Biophys.*, **261**, 175–81.

Jarrett, L. and Seals, J.R. (1979). Pyruvate dehydrogenase activation in adipocylte mitochondria by an insulin-generated mediator from muscle. *Science*, **206**, 1407–8.

Kiechle, F.L., Jarrett, L., Kotagal, N. and Popp, D.A. (1981). Partial purification from rat adipocyte plasma membranes of a chemical mediator which stimulates the action of insulin on pyruvate dehydrogenase. *J. Biol. Chem*, **256**, 2945–51.

Larner, J. (1988). The rat liver insulin-mediator contains galactosamine and chiro-inositol. *Biochem. Biophys. Res. Commun.*, **151**, 1416–26.

Low, M.G. (1989). Glucosyl-phosphatidylinositol: a versatile anchor for cell surface proteins. *FASEB J.*, **3**, 1600–8.

Mato, J.M., Kelly, K.L., Abler, A. *et al.* (1987). Partial structure of an unsulin-sensitive glycophospholipid. *Biochem. Biophys. Res. Commun.*, **146**, 764–70.

Romero, G. (1988). Inositol phosphate glycan anchors of membrane proteins as potential precursors of insulin mediators. *Science*, **240**, 509–12.

Saltiel, A.R. (1988). In search of a second messenger for insulin. *Amer. J. Physiol.*, **255**, C1–C11.

Saltiel, A.R., Fox, J.A., Sherline, P. *et al.* (1986). Insulin stimulated hydrolysis of a novel glycolipid generates modulators of cAMP phosphodiesterase. *Science*, **233**, 967–72.

Saltiel, A., Jacobs, S., Siegel, M. *et al.* (1981). Insulin stimulates the release from liver plasma membranes of a chemical modulator of pyruvate dehydrogenase. *Biochem. Biophys. Res. Commun.*, **102**, 1041–7.

Suzuki, K. and Kono, T. (1980). Evidence that insulin causes translocation of glucose transport activity to the plasma membrane from an intracellular storage site. *Proc. Natl. Acad. Sci. USA*, **77**, 2542–5.

Wilson, S.R., Wallace, A.V. and Houslay, M.D. (1983). Insulin activates the plasma-membrane and dense-vesicle cyclic AMP phosphodiesterase in hepatocytes by distinct routes. *Biochem. J.*, **216**, 245–8.

Polypeptide growth factors

Carpenter, G. (1987). Receptors for epidermal growth factor and other polypeptide mitogens. *Ann. Rev. Biochem.*, **56**, 881–914.

Carpenter, G. and Cohen, S. (1984). Peptide growth factors. *Trends Biochem. Sci.*, **9**, 169–272.

Chapron, Y. Cochet, C. Crouzy, S. *et al.* (1989). Tyrosine protein kinase activity of the EGF receptor is required to induce activation of receptor-operated calcium channels. *Biochem. Biophys. Res. Commun.*, **158**, 527–33.

Chen, W.S., Lazar, C.S., Poenie, M. *et al.* (1987). Requirement for intrinsic protein tyrosine kinase in the immediate and late actions of the EGF receptor. *Nature*, **328**, 820–3.

Cohen, S. (1986). Epidermal growth factor. *Biosci. Rep.*, **6**, 1017–28.

Gill, G.N., Bertics, P.J. and Santon, J.B. (1987). Epidermal growth factor and its receptor. *Mol. Cell Endocrinol.*, **51**, 169–86.

James, R. and Bradshaw, R. (1984). Polypeptide growth factors. *Ann. Rev. Biochem.*, **53**, 259–92.

Newmark, P. (1987). Oncogenes and cell growth. *Nature*, **327**, 101–2.

Perdue, J.F. (1984). Chemistry, structure, and function of insulin-like growth factors and their receptors: a review. *Can. J. Biochem. Cell Biol.*, **62**, 1237–45.

Schlessinger, J. (1988). The EGF-receptor as a multifunctional allosteric protein. *Biochemistry*, **27**, 3113–18.

Williams, L.T. (1989). Signal transduction by the platelet-derived growth factor receptor. *Science*, **243**, 1564–70.

Yarden, Y. (1988). Growth factor receptor tyrosine kinases. *Ann. Rev. Biochem.*, **57**, 443–78.

7 Arachidonic acid and eicosanoids

Introduction

The existence of lipid-soluble 'hormone'-like agents which are active *in vitro* was first demonstrated almost 60 years ago when it was observed that seminal fluid from several species of mammal contains an active principle which can induce smooth-muscle contraction. It was later found that this substance is not a single molecular species but that such lipid extracts contain several related compounds which are all composed of 20 carbon atoms and possess a terminal carboxyl group. The compounds were called 'prostaglandins' to denote their supposed source of origin in the prostate gland. This has proved, however, to be an inappropriate nomenclature since many (if not all!) tissues are capable of synthesizing such molecules, and they are known to carry out important messenger functions at a local level. Originally, it was proposed that prostaglandin products are derived from a hypothetical precursor termed 'prostanoic acid' which was envisaged to contain 20 carbon atoms and a carboxyl group but no unsaturated double bonds. Subsequently, it became clear that this is not the case but that the precursor is a fatty acid which is an essential component of the human diet, 5,8,11,14-eicosatetraenoic acid, or more commonly, arachidonic acid (Fig. 7.1). It has emerged in subsequent years, that arachidonic acid can be metabolized by several different routes in cells to yield a very large number of pharmacologically-distinct products. These include thromboxanes, leukotrienes and prostacyclins in addition to the originally discovered prostaglandins and are now referred to under the more general collective heading of 'eicosanoids'. None of these compounds is stored in cells, which means that they must be synthesized *de novo* from the precursor arachidonic acid immediately upon cell stimulation. Indeed, it seems that the major determinant of their rate of synthesis is the availability of free arachidonic acid. Normally, this molecule is not present in cells in a free form but is esterified within one or more of the lipid components of the cell. In this esterified form, arachidonic acid is not available to the synthetic enzymes responsible for eicosanoid synthesis and must first be released from its site of sequestration by a specific hydrolase enzyme. Arachidonic acid is present in a range of lipid molecules, including tri- and di-acylglycerols, cholesterol esters and phospholipids. It is the last class of molecules that represents the major source of readily mobilizable arachidonic acid, although, under certain circumstances, diacylglycerol can also serve as the source. In general, arachidonic acid is most commonly found esterified at position 2 of the

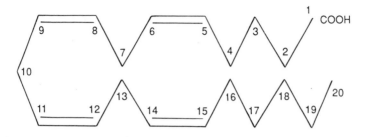

Fig. 7.1   Structure of 5,8,11,14-eicosatetraenoic acid (arachidonic acid)

glycerol backbone of cellular phospholipids and is, therefore, particularly susceptible to hydrolysis by phospholipases (PL) of the 'A$_2$' subtype which preferentially cleave fatty acids located at this position. It is, therefore, the regulation of PLA$_2$ activity which directly controls the release of arachidonic acid in most cells and it is now clear that this enzyme is subject to rigorous control in the cell.

Phospholipase A$_2$

PLA$_2$ activity has been observed in a diverse array of tissues in organisms ranging from bacteria to man, and as many as 40 separate enzyme activities have been defined. These share a number of common characteristics, including heat and acid stability and sensitivity to Ca^{2+} ions. In general, the enzymes are small proteins (approximate molecular weights 15 K) which show some degree of structural homology at the amino acid level. It should be emphasized, however, that PLA$_2$ subserves a number of different functional roles which can be either intra- or extra-cellular according to circumstances, and although the different species of enzyme exhibit similarities, there is no reason to suppose that they will all be subject to identical control mechanisms. For example, PLA$_2$ is a prominent component of some snake venoms and is also released into the gastrointestinal tract from the pancreas. It is clear that its intended role in these situations cannot be equivalent to its regulatory function in the intracellular signalling mechanisms of cells and it is likely that the enzyme activity will be controlled by different factors in each specific circumstance. For this reason, it is necessary to exercise due caution when studying the regulation of PLA$_2$ activity *in vitro*. For example, many studies have used the snake venom or pancreatic enzyme for investigation of the regulation of PLA$_2$, when it is by no means certain that these particular forms of the enzyme are controlled in an equivalent manner to those present intracellularly. For the purposes of the present account it is most appropriate to focus on the regulation of PLA$_2$ activity inside cells since it is this activity which largely determines the availability of free arachidonic acid. In another note of caution, it is also worth emphasizing that PLA$_2$ activation is often measured (in intact

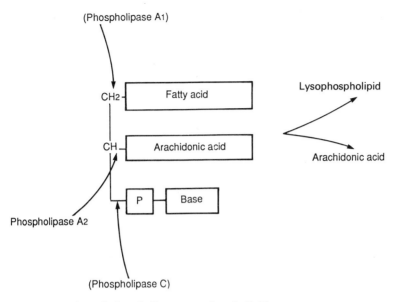

Fig. 7.2 Action of phospholipases on phospholipids

cells) by monitoring the release of radiolabelled arachidonic acid from pre-labelled lipid fractions. While this can certainly be indicative of PLA_2 activity, it is not an absolute measure, since other enzymes (e.g. DG lipase) can also elicit arachidonic acid release under appropriate circumstances.

The PLA_2 responsible for mobilizing arachidonic acid from membrane lipids is likely to be predominantly localized to the particulate fraction of cells, although it is apparently not entirely confined to the plasma membrane since it has also been found in the membranes of cellular organelles, especially secretory granules. The enzyme from most tissues requires Ca^{2+} for activity and a defined Ca^{2+}-binding site has been identified in the enzyme structure. This suggests that the availability of Ca^{2+} could play a significant role in controlling enzyme activity and it has indeed, been shown that PLA_2 can be activated in tissue extracts by incubation with increasing concentrations of Ca^{2+}. Some results have implicated calmodulin in mediation of this effect, although the presence of a Ca^{2+}-binding site on the enzyme itself suggests that a mediator protein may not be required. Indeed, isolation and characterization of the membrane-associated PLA_2 activity from platelets has provided no evidence of a role for calmodulin in the activation process. One problem with a model which envisages the concentration of intracellular Ca^{2+} as a regulator of PLA_2 is that the enzyme displays unusually low sensitivity to Ca^{2+} in *in vitro* assay systems (Fig. 7.3) Thus, it is often necessary to increase the prevailing Ca^{2+} concentration into the sub-millimolar range in order to elicit full enzyme activity. Such concentrations are extremely non-physiological and are unlikely to prevail in intact cells, even at sites of restricted Ca^{2+} diffusion, such as may occur close to the plasma membrane.

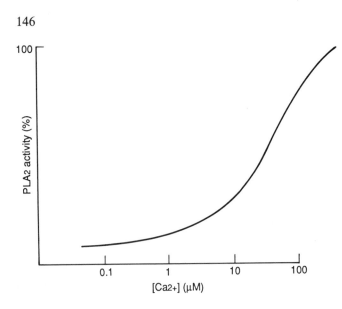

Fig. 7.3 Sensitivity of phospholipase A_2 to Ca^{2+} ions

Thus, although it seems likely, it is still not certain that PLA_2 activation is directly mediated by a change in cytosolic Ca^{2+} concentration. In at least one tissue (the insulin producing B-cells of islets of Langerhans) it has been reported that cell stimulation (with glucose) is associated with a stable activation of PLA_2 that persists through tissue homogenization. Such an effect could not be directly achieved by a change in Ca^{2+} concentration, since activation by this mechanism would be lost as soon as the tissue is disrupted. If confirmed for other cell types, this might suggest that PLA_2 is subject to covalent modification (phosphorylation perhaps?) as part of its mechanism of regulation.

More recently, the possible intervention of a G-protein in the agonist-induced activation of PLA_2 has been considered. For example, it has been shown that PLA_2 activity in permeabilized thyroid cells can be stimulated by introduction of the non-hydrolysable guanine-nucleotide analogue GTP-γ-S, and that this response is susceptible to inhibition by pertussis toxin. These results are highly suggestive of G-protein involvement. The nature of this putative transducer protein has not been determined, although studies performed with cells obtained from vertebrate retina have shown that PLA_2 activity can be increased under conditions favouring activation of the G-protein transducin. Surprisingly, this response correlated more closely with the release of the $\beta\gamma$ subunits of transducin than with the generation of free α-subunit. If this observation is corroborated by experiments in other tissues, it may pave the way to the elucidation of a novel regulatory mechanism involving the 'inactive' subunits of G-proteins. Furthermore, since the $\beta\gamma$ complex is structurally very similar in all known G-proteins, this mechanism could be operative in other cells which lack the α-subunit of transducin. Other factors must also be important, however, otherwise PLA_2 would be indiscriminately activated by all agents causing G-protein dissociation. Fortunately this does not occur!

Lipocortin as a regulator of phospholipase A_2

The possibility that endogenous PLA_2 could be subject to direct regulation by a cellular protein was first suggested by the observation that treatment of neutrophils with synthetic chemoattractants release arachidonic acid from membrane phospholipids, and that this effect can be inhibited by prior incubation of the cells with glucocorticoid hormones. Since glucocorticoids modulate the state of cell activation by inducing new protein synthesis, the concept emerged that they might be directly controlling the production of a modulator protein which itself controls the rate of arachidonic acid hydrolysis from phospholipid. This idea was also supported by experiments showing that the anti-inflammatory properties of steroids correlate with decreased production of inflammatory mediators derived from arachidonic acid, such as prostaglandins and leukotrienes. Since these agents arise as a result of the metabolism of arachidonic acid, this also suggests that there is a primary control mechanism regulating arachidonic acid availability in steroid treated cells. On the basis of such studies, a protein was identified in cell extracts that would inhibit PLA_2 activity in *in vitro* assay systems. This protein (or a very similar one) was observed in several cells including macrophages, neutrophils, adrenal medulla and platelets and was characterized as a species of about 40-K molecular weight. Originally, the protein was given a variety of names which reflected its various tissues of origin, and these include 'lipomodulin', 'macrocortin' and 'renocortin'. More recently, a standardized nomenclature has been adopted and the protein is now generally referred to as 'lipocortin'. Indeed, it has since been established that a family of lipocortin-like proteins exist and it seems likely that individual cells probably express either a particular species of protein or a combination of such proteins which allows PLA_2 activity to be controlled in an appropriate fashion according to the type of hormonal signals which impinge upon that cell. In this context it is worth noting that the dose–response curve for induction of lipocortin synthesis by glucocorticoid hormones is such that synthesis will be slightly stimulated even at normal circulating concentrations of the hormones. Thus, many steroid-responsive tissues (i.e. most cells in the body) are likely to express some lipocortin under basal conditions, which implies that PLA_2 activity could be normally repressed in cells because of the endogenous production of lipocortin. This, in turn, indicates that there are likely to be other regulatory mechanisms at work which permit the acute regulation of PLA_2 activity following cell stimulation.

The gene for one form of lipocortin has recently been cloned and expressed in *Escherichia coli* which has allowed the first detailed structural analysis of this molecule. In these experiments, lipocortin was first purified from rat peritoneal tissue and used to obtain amino acid sequence data, which then permitted the synthesis of oligonucleotide hybridization probes which could be radiolabelled with ^{32}P. These were found to recognize RNA molecules from human liver suggesting that rat and human lipocortins are structurally very similar. A cDNA library was subsequently constructed and screened and the lipocortin gene was eventually isolated and cloned. The gene codes for a protein of 346 amino acids which, when isolated, is active as an inhibitor of PLA$_2$ *in vitro*. Interestingly, analysis of human DNA suggested that there is only one gene for this particular type of lipocortin which has now been called 'lipocortin I', indicating that other members of this family of proteins must either be derived from an original mRNA transcript by differences in processing or protein structural organization (e.g. proteolytic cleavage or polymer formation), or that the other PLA$_2$ inhibitory proteins are structurally different from this lipocortin. In this context, a second lipocortin gene has also been sequenced which codes for a protein that differs from the first at approximately 50% of the amino acid residues. This second form of lipocortin has been designated type II and the two forms are clearly entirely separate proteins. They do, however, share certain common structural features and both are effective inhibitors of PLA$_2$. The significance of the existence of two isoforms of the molecule may lie in differential regulatory mechanisms as outlined below.

Human lipocortin I is a relatively hydrophilic molecule which contains a potential glycosylation site near to the amino terminus of the molecule. Normally, it is located intracellularly and, since it is very polar, it presumably resides in the cytosol of the cell. This suggests that its interaction with PLA$_2$ is also likely to occur in the cytosolic domain despite the clear implication that the biological activity of PLA$_2$ is expressed at the level of cell membranes. Hence, this is another example of a cytosolic interaction which controls the rate of membrane-associated events.

A second feature of the lipocortin molecule revealed by structural analysis is that it contains two sequences of amino acids which represent consensus sequences for phosphorylation by two protein kinases, PK-A and PK-C. This is of interest since evidence has been provided which demonstrates that the ability of lipocortin to inhibit PLA$_2$ can be regulated by protein kinase activation. In particular, it has been shown that both PK-A and PK-C can phosphorylate the purified molecule *in vitro* and that activation of these kinases in intact cells leads to phosphorylation of a protein that is precipitated by antibodies raised against lipocortin. Significantly, the phosphorylated form of lipocortin is much less effective as an inhibitor of PLA$_2$ than is the native molecule. Therefore, in intact cells, activation of PK-A or PK-C (and probably other Ca^{2+}-dependent protein kinases as well) will lead to phosphorylation of lipocortin, relief of inhibition of PLA$_2$ and a consequent increase in the rate of release of arachidonic acid from membrane phospholipids. By this means hormones acting on other second messenger systems could indirectly regulate the availability of free arachidonic acid in cells and thereby control apparently independent pathways by activation of a single protein kinase (Fig. 7.4). If inhibition of the restraining influence of lipocortin is also accompanied by conditions which will directly favour PLA$_2$ activation (e.g. a rise in cytosolic

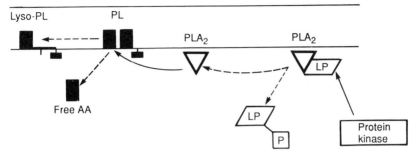

Fig. 7.4 Regulation of phospholipase A_2 activity by lipocortin. Lyso-PL = lysophospholipid; PL = phospholipid; PLA_2 = phospholipase A_2; AA = arachidonic acid; LP = lipocortin

Ca^{2+} concentration) it can easily be envisaged that amplification of a hormonal signal will result.

A second very interesting observation is that lipocortin can also serve as a substrate for some tyrosine specific protein kinases. The target phosphorylation site for these enzymes is located near to the N-terminus of the molecule in a region which varies between lipocortin types I and II. Hence, the two isoforms could possibly be subject to independent regulation by distinct tyrosine kinase enzymes. Indeed, it appears that lipocortin I is a good substrate for the tyrosine kinase associated with the EGF-receptor, while lipocortin II is preferentially phosphorylated by a tyrosine kinase enzyme encoded by the Rous sarcoma virus (known as pp60^{v-src}). These observations raise the possibility that the two types of lipocortin may be subject to differential regulation in intact cells and that a degree of specificity is introduced into the cellular regulation of PLA_2 activity according to the relative abundance of each type of lipocortin.

It is not yet entirely clear how lipocortin exerts its influence on PLA_2 activity, but it has been demonstrated that under optimal conditions maximal inhibition can be achieved in the presence of approximately equal stoichiometric amounts of both proteins. This suggests that a 1:1 ratio of inhibitor to enzyme is necessary and may imply a direct interaction between the two molecules. Kinetic experiments have been interpreted as indicating that the primary effect of lipocortin is to decrease the V_{max} of PLA_2 for substrate phospholipids while leaving the K_m unchanged.

An alternative mechanism that has been proposed to explain the inhibitory effect of lipocortin on PLA_2 activity stems from the observation that lipocortin has several binding sites for phospholipid. Thus, it is possible that the regulator protein could directly interact with lipid molecules located in the membrane and, in so doing, lead to a change in the disposition of these lipids such that they no longer act as effective substrates for PLA_2. This indirect inhibitory mechanism has seemed attractive since, in some studies, it has been necessary to use excess lipocortin in order to obtain effective inhibition of PLA_2. However, such results have often derived from *in vitro* experiments in which the source of PLA_2 was different from that of the lipocortin. In particular, use has often been made of either the snake venom or pancreatic PLA_2, both of which have extracellular targets and are unlikely to be as tightly controlled as the intracellular enzyme. It remains uncertain, therefore, whether the effects of lipocortin on such forms of PLA_2 are identical to those which cause inhibition of endogenous PLA_2 in cells.

Firm evidence that direct interactions can indeed occur between PLA_2 and lipocortin has been provided by the demonstration that the enzyme can be purified from cell extracts by application to a lipocortin affinity column. PLA_2 readily absorbs to the column, suggesting that it must be able to interact with lipocortin with high affinity. It seems likely that both lipocortin I and lipocortin II share similar mechanisms for mediating inhibition of PLA_2 since proteolytic fragments have been used to define regions of lipocortin I that are responsible for enzyme inhibition, and the most active fragment corresponds to a region which is highly conserved between the two molecules.

Arachidonic acid

Alternative pathways for arachidonic acid release

The major substrate(s) from which arachidonic acid is released following activation of PLA_2 varies from cell to cell, although phosphatidylcholine and phosphatidylethanolamine seem to be rich sources of arachidonic acid in many tissues. However, another phospholipid pool which is often enriched in arachidonic acid is the metabolically-active inositol-containing phospholipid pool. These lipids play a unique role in cell Ca^{2+} homeostasis (see Chapter 5) but can also serve as a precursor pool of readily-mobilizable arachidonic acid. Release of arachidonic acid from phosphatidylinositol and its phosphorylated derivatives occurs following an initial hydrolysis of the lipids by phospholipase C (PL-C), which suggests that the actual substrate from which arachidonic acid is derived is not the phospholipid itself but one of the products of hydrolysis, namely diacylglycerol (DG). This reaction is catalysed by a specific DG-lipase which does not use phospholipid as substrate but which, like PLA_2, preferentially hydrolyses the fatty acid esterified to carbon-2 of the glycerol backbone. Hence, the concerted action of PL-C and DG-lipase can elicit arachidonic acid release in cells undergoing rapid phospholipid turnover. This mechanism will not be operative in resting cells, however, since endogenous DG levels are very low under such conditions.

Arachidonic acid as a second messenger

In many cells the release of free arachidonic acid from membrane lipids is the first step in the generation of a cascade of potential messengers. This is because arachidonic acid can be metabolized by several different enzyme systems (see below) to yield a wide range of metabolites. Some of these act as intracellular messengers by direct actions on target enzymes (which remain largely undefined) while many are released into the extracellular environment where they bind to, and activate, surface receptors on neighbouring cells. However, it is emerging that arachidonic acid itself may serve as a messenger molecule in some cells, without the requirement for further metabolism. For example, under certain conditions arachidonic acid can elicit the release of hormones, such as prolactin and insulin, when it is directly added to the incubation medium bathing cells. These effects are not blocked by inhibitors of arachidonic acid metabolism which suggests that they probably reflect direct actions of the fatty acid itself.

Recently, at least two possible intracellular mechanisms have been revealed which might contribute to the direct effects of arachidonic acid in cells. The first of these involves activation of one or more protein kinase enzymes. In particular, it has been shown that several long-chain fatty acids (including arachidonic acid) can promote the activation of a calcium-dependent protein kinase, protein kinase C (PK-C) (Chapter 8). The effect of arachidonic acid is to increase the sensitivity of this enzyme for Ca^{2+}, and thus to promote activation of PK-C at lower levels of Ca^{2+} than would normally be required. The K_m for activation by arachidonic acid is approximately $15\,\mu M$ and, surprisingly, higher concentrations of the fatty acid actually inhibit protein kinase activity in some cells. However, this second response is only seen at very high concentrations of arachidonic acid ($>100\mu M$), suggesting that it is unlikely to be of any physiological significance.

A second response elicited by the fatty acid involves the release of Ca^{2+} from intracellular storage sites. This effect has been observed in pituitary cells, liver and pancreatic islets, although the precise characteristics seem to vary from one cell to another. Hence, in insulin-secreting cells, arachidonic acid causes the release of Ca^{2+} from a pool which is different from that normally mobilized by stimuli which activate inositol lipid hydrolysis (Chapter 5), whereas in pituitary cells, the two mechanisms seem to induce the release of Ca^{2+} from the same pool. The potential significance of these effects is obvious, since Ca^{2+} is a regulator of many cellular processes, including secretion. However, the ability of arachidonic acid to mobilize intracellular Ca^{2+} has largely been observed either in isolated subcellular fractions or in cells which have been treated with detergents to render them freely permeable to the fatty acid. It has proved much more difficult to observe arachidonic acid-induced Ca^{2+} release in intact cells even when conditions are employed which favour activation of phospholipase A_2 and release of endogenous arachidonic acid. This does not exclude this mechanism as a possible mediator of the effects of arachidonic acid in cells, but it does imply that further research is still necessary to clarify its importance. In this context, some evidence suggests that arachidonic acid can actually promote Ca^{2+} extrusion from stimulated cells and that it inhibits the increase in intracellular Ca^{2+} concentration brought about by agents that activate inositol lipid turnover. This is the situation, for example, in a cultured line of pituitary cells (GH$_3$) which respond to thyrotropin-releasing hormone (TRH) with the release of intracellular Ca^{2+} and secretion of prolactin. Arachidonic acid lowers the cytosolic Ca^{2+}

concentration under these conditions and promotes the efflux of Ca^{2+} from the cells.

Hence, there is accumulating evidence to support a role for arachidonic acid as an intracellular messenger molecule, but the details of its effects remain to be firmly established and may even vary from tissue to tissue.

Metabolism of arachidonic acid

The first products of arachidonic acid metabolism to be isolated and characterized from tissue extracts were prostaglandin compounds which were coined prostaglandins E (PGE) and F (PGF). The nomenclature seems rather unusual but was derived from the observed solubility of these compounds in either ether (PGE) or phosphate buffers (PGF-F for phosphate!). It was subsequently observed that these molecules can be synthesized from free arachidonic acid by tissue homogenates in a reaction catalysed by the rate-limiting enzyme, fatty acid cyclooxygenase. More recently, alternative routes of arachidonic acid metabolism have also been identified, and rely on an initial hydroperosixe formation catalysed by one of several lipoxygenase enzymes. All of these routes probably give rise to biologically important messengers and it will be appropriate to briefly consider each pathway in turn. It is important to remember, however, that not all cells contain all of the enzymes capable of metabolizing arachidonic acid and that the particular spectrum of products generated is likely to be tissue-specific.

Metabolism of arachidonic acid by the cyclooxygenase pathway

When PGs were first discovered, their synthesis was attributed to the activity of a hypothetical enzyme, dubbed 'prostaglandin synthase'. Subsequently, study of the metabolic pathways leading to PG formation revealed that the initial reaction products are unstable molecules containing peroxide moieties within their structures, and the enzyme responsible for their synthesis has now been renamed to take account of this mechanism – it is therefore, prostaglandin endoperoxide synthase (PES). The important feature of this enzyme is that it catalyses the insertion of two molecules of oxygen into arachidonic acid to yield an endoperoxide derivative (a 15-hydroperoxy-9, 11-endoperoxide) and then the same protein reduces the newly formed peroxide at position 15 to a hydroxyl derivative. Thus, the enzyme possesses two separate catalytic activities (a cyclooxygenase and a peroxidase) which combine to produce a trioxygenase derivative of arachidonic acid, PGH_2 (Fig. 7.5). This molecule then serves as the immediate precursor for various other enzymes which give rise to the diversity of PG products.

PES has been identified in a wide range of different tissues and has now been purified to homogeneity. The enzyme requires detergent for purification, suggesting that it exists in a membrane-associated form (probably largely on the endoplasmic reticulum) and it contains carbohydrate residues. When purified in native form the enzyme appears to exist as a dimer, but

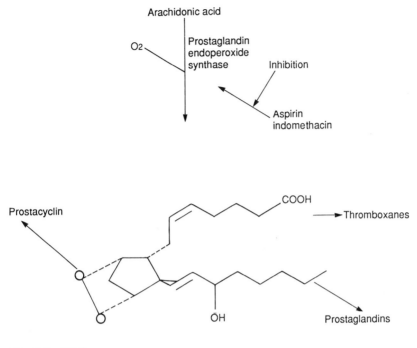

Fig. 7.5   PGH$_2$

polyacrylamide gel electrophoresis under reducing conditions has revealed only a single type of subunit ($M_r \sim 72$ K). Each molecule of subunit contains a haem prosthetic group which is essential for enzyme activity, and the enzyme displays high affinity ($K_m \sim 0.5 \, \mu$M) for both of its substrates (arachidonic acid and O$_2$). One unusual aspect of the chemistry of PES is that its cyclooxygenase activity displays self-inhibitory properties, such that in the presence of both substrates, the enzyme consumes O$_2$ more rapidly than arachidonic acid, and then becomes inactive. Some evidence suggests that this mechanism may not be a unique property of the isolated enzyme but that it also occurs in intact cells. If so, then this could clearly be a limiting factor in PG production. Surprisingly, perhaps, the cyclooxygenase activity can be maintained if the enzyme is supplied with exogenous hydroperoxides, an observation which has promoted the idea that a certain basal level of peroxide formation in cells may be necessary to maintain PES in an active state.

PES represents the target site for one of the most widely used of all drugs, aspirin, which is an effective inhibitor of PG synthesis. This response results from acetylation of the enzyme in the presence of aspirin, leading to irreversible inactivation. Another widely-used PES inhibitor is indomethacin which, like aspirin, selectively affects prostaglandin formation while leaving production of other eicosanoids unaltered (Fig. 7.5).

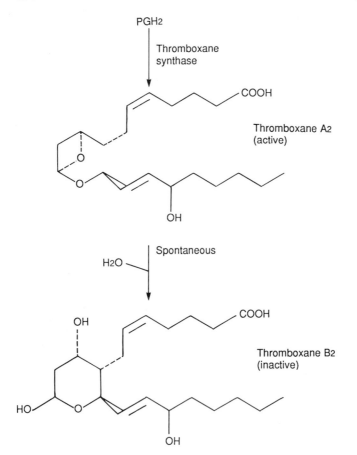

Fig. 7.6   Synthesis of thromboxanes from PGH_2

Endoperoxide metabolism

Thromboxanes
Studies of the metabolism of PGs in platelets led to the observation that one of the products derived from the unstable precursor PGH_2 contains an oxane ring and a hemi-acetal hydroxyl residue. This molecule has been termed thromboxane B_2 (TxB_2) but is inactive as a vasoconstrictor and as an inducer of platelet activation. It was subsequently established that TxB_2 represents the spontaneous degradation product of another, biologically active, PG derivative which contains an oxetane ring and is referred to as TxA_2 (Fig. 7.6). This molecule has an extremely short biological half-life (~ 30 sec) but is a potent vasoconstrictor and platelet activator (active in the nanomolar range). TxA_2 is readily synthesized by platelets but may also be produced in other tissues, and is generated by a membrane-associated enzyme whose subcellular distribution

parallels that of PES. Thromboxane synthase is active over a wide pH range (pH 5–9) and has been partially purified by affinity chromatographic methods. The purified enzyme has certain properties in common with cytochrome P_{450} and may be a similar haem-containing enzyme.

Prostacyclin

In the cells of the vascular epithelium, PGH_2 is not metabolized to yield thromboxanes, as in platelets, but rather is converted to a reactive species which has properties that are antagonistic to those of TxA_2. This molecule bears an enol-ether moiety in its structure which gives it a cyclic configuration and, hence, it is often referred to as prostacyclin (PGI_2) (Fig. 7.7). PGI_2 is a vasodilator and has a biological half-life of ~ 10 min at physiological pH although it is much more stable in alkaline solutions (half-life of several days). Decomposition of PGI_2 results in the formation of a stable end-product, 6-

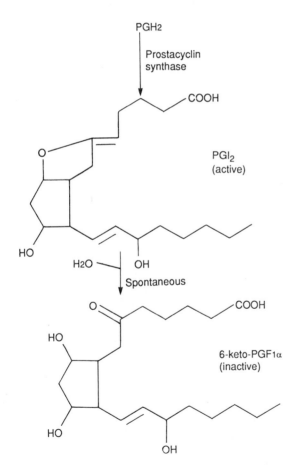

Fig. 7.7 Synthesis of prostacyclin (PGI_2) from PGH_2

keto-PGF$_{1\alpha}$ which can be conveniently measured by radioimmunoassay, but is biologically non-functional.

PGI$_2$ is formed from PGH$_2$ by the action of prostacyclin synthase which has now been purified in homogeneous form. The enzyme has been localized by immunological staining techniques and seems to be most abundant in the plasma membrane and nuclear membrane of vascular smooth muscle. It is apparently also present in other tissues, although it can be difficult, using biochemical methods, to conclusively demonstrate that the enzyme is located in the cells of any particular tissue, since contaminating blood vessels can often account for much of the measured activity. Purified prostacyclin synthase has a high affinity for PGH$_2$ ($K_m \sim 5\,\mu$M) and it displays optical properties consistent with it being a cytochrome P$_{450}$-type enzyme. The subunit molecular weight is 50 K on polyacrylamide gels and the enzyme displays an autocatalytic inactivation process. The activity may also be sensitive to free-radical production (leading to inhibition) and this has been postulated to explain the mechanism by which prostacyclin synthase is often observed to become inactivated during tissue culture of cells. It has also been suggested that such regulation of enzyme activity could explain the observation that PGI$_2$ production sometimes occurs in a phasic pattern *in vivo*. However, this could also reflect rapid inactivation of cyclooxygenase rather than a direct effect on prostacyclin synthase *per se*.

The metabolic responses to PGI$_2$ production seem to reflect increased cell cAMP levels and a receptor protein for the compound has been identified on responsive cells.

This receptor is located at the cell surface and is coupled to adenylate cyclase in a stimulatory manner. Thus, PGI$_2$ acts as a local 'hormone', being produced in one tissue in response to cell stimulation and then directly controlling the activity of another cell type which is located in the immediate vicinity.

Prostaglandins D$_2$ and E$_2$

PGD$_2$ is formed from PGH$_2$ by an isomerization reaction which can be catalyzed by several apparently different enzymes (Fig. 7.8). The most unlikely of these is serum albumin which exhibits a K_m for PGH$_2$ of $6\,\mu$M. This emphasizes the need for extreme caution when attempting to quantitate the PGD$_2$ synthetic pathway in tissue extracts since buffer composition can dramatically affect any measured reaction velocities. Despite this (presumably) non-specific effect of albumin, at least two other enzymes have been identified which catalyse the isomerization of PGH$_2$. One of these is found in brain ($M_r \sim 85$ K) and another in spleen ($M_r \sim 30$ K). The enzymes have different properties from each other and both display specific activities which are more than 100 times that of albumin.

PGD$_2$ acts as an antagonist to TxA$_2$ and, like PGI$_2$, binds to a cell surface receptor that is positively coupled to adenylate cyclase. This receptor is not the same as that for PGI$_2$, and appears to be expressed in high levels on neuronal cells. This may point to a role for PGD$_2$ in neuromodulation.

In some cells, notably kidney, PGH$_2$ also acts as substrate for a separate

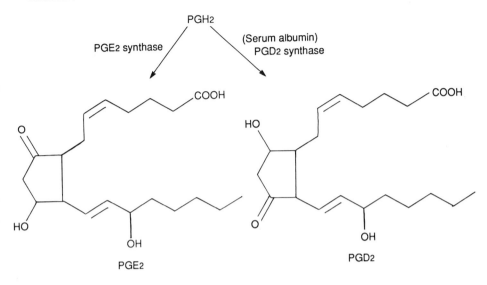

Fig. 7.8 Isomerization of PGH_2

isomerase, which converts it to PGE_2, a metabolite that acts antagonistically to vasopressin and promotes urinary water loss. PGE_2 synthase has not yet been obtained in pure form but, like many other PG synthetic enzymes, it also seems to be membrane-associated.

Metabolism of arachidonic acid by the lipoxygenase pathway

Unlike cyclooxygenases which catalyse the incorporation of two molecules of O_2 into fatty acids, lipoxygenase enzymes catalyse the transfer of only one O_2 molecule to yield a hydroperoxy derivative. Lipoxygenases have been known for more than 50 years as plant enzymes but it is only within the last 15 years that their importance in animal tissues has been recognized. They act exclusively on unsaturated fatty acids and catalyse the formation of a conjugated diene bearing the hydroperoxide group. The enzymes exhibit marked stereospecificity and are also specific with respect to the position in the fatty acid chain at which they introduce molecular oxygen. Three distinct types of lipoxygenase enzyme have, so far, been identified in mammalian cells: arachidonic acid 12-lipoxygenase; arachidonic acid 15-lipoxygenase; arachidonic acid 5-lipoxygenase.

Arachidonic acid 12-lipoxygenase

The existence of this enzyme in mammals was first postulated, following the observation that platelets could convert arachidonic acid into 12-hydroxy-eicosa-5,8,10,14-tetraenoic acid (12-HETE) in a manner which was not inhibited by either aspirin or indomethacin, both of which are powerful inhibitors of arachidonic acid metabolism via the cyclooxygenase pathway

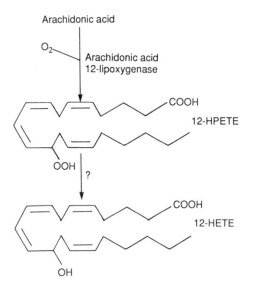

Arachidonic acid

Arachidonic acid
12-lipoxygenase

12-HPETE

?

12-HETE

Fig. 7.9 Formation of 12-HETE from arachidonic acid

(Fig. 7.9). 12-HETE is a relatively stable molecule and was viewed as the probable breakdown product of a more primary product which bears a hydroperoxy group in place of the hydroxyl moiety, i.e. 12-hydroperoxy-eicosa-5,8,10,14-tetraenoic acid (12-HPETE). This was confirmed by the subsequent isolation of 12-HPETE as the reaction product of platelet lipoxygenase, although it is clear that 12-HPETE is rapidly converted to 12-HETE in many biological tissues. It is not entirely certain how this transformation occurs although an enzymatic conversion seems most likely. Indeed, the enzyme glutathione reductase has been implicated in the process, although some evidence suggests that the 12-lipoxygenase itself may be capable of catalysing the reaction.

Arachidonic acid 12-lipoxygenase has been characterized in blood platelets where it exists in both particulate and soluble forms. The latter, however, represents an aggregated form of the enzyme which exists in a complex that also contains lipid. The two types of enzyme (soluble and particulate) cannot be distinguished on the basis of any physicochemical properties, suggesting that they probably represent two forms of the same molecule. Recent estimates suggest that the 12-lipoxygenase is a large protein and species with molecular weights of both ~ 100 K and 160 K have been identified with the enzyme. It is possible that the larger of these two forms may represent a complex containing both the 12-lipoxygenase and an attendant peroxidase activity, but further studies are required to unequivocally resolve this. 12-Lipoxygenase has a K_m for arachidonic acid of $\sim 5\,\mu$M and has not been shown to have any specific requirement for co-factors.

Although 12-lipoxygenase activity has been found in platelets from all species examined (including man), the role of the products of this enzyme remains a mystery. Platelet-derived 12-HPETE may play a role in neutrophil activation by stimulating the metabolism of arachidonic acid in these cells, and the endogenous activity of 12-lipoxygenase has been proposed as an important component of the stimulus–secretion coupling mechanism of the insulin secreting B-cells of islets of Langerhans. However, the mechanistic details and true biological significance of these processes have not been established.

Arachidonic acid 15-lipoxygenase
Neutrophils and reticulocytes possess an arachidonic acid 15-lipoxygenase which is a soluble enzyme of approximately 70 K molecular weight. It has a relatively high K_m for arachidonate ($\sim 30 \mu M$) which requires that very substantial amounts of the fatty acid must be released from membrane lipids in order to promote significant conversion to 15-HPETE. Interestingly, studies with reticulocytes suggest that the 15-lipoxygenase enzyme present in these cells may be capable of utilizing esterified arachidonic acid as substrate, thus providing the capacity to modify the membrane environment *in situ* without the necessity to mobilize the arachidonic acid in free form. In cells which produce 15-HPETE, spontaneous degradation appears to occur with the resultant generation of 15-HETE together with several other products in varying proportions. Some of these products can then serve as substrates for a 5-lipoxygenase enzyme and, hence, the 15- and 5-lipoxygenases may act in concert to yield a variey of potential messenger substances.

It has been argued that 15-HETE could itself control the relative distribution of other lipoxygenase enzymes in platelets and, conversely, to activate a 5-lipoxygenase activity in cultured mast cells. Much more work is still required, however, in order to firmly establish the basis and significance of these effects.

Arachidonic acid 5-lipoxygenase
The products of the arachidonic acid 5-lipoxygenase have been known for more than 40 years as potent bronchoconstrictors which were collectively referred to as 'slow-reacting substance of anaphylaxis' (SRS-A). The chemical structure of these compounds and the route of their formation has only emerged in the past 10 years or so, since it was established that some cells can metabolize arachidonic acid via a 5-lipoxygenase enzyme to yield a family of biologically-active molecules that are now referred to as leukotrienes. The first evidence for the existence of this pathway came from neutrophils where it was observed that the major product derived from arachidonic acid is 5-HETE. It has since been established that this molecule is derived from 5-HPETE which can also be transformed to a series of more polar species in order to give rise to the leukotrienes. The 5-lipoxygenase enzyme responsible for synthesis of 5-HPETE from arachidonic acid has now been purified and is composed of a single polypeptide chain ($M_r \sim 80$ K). The enzyme appears to possess two distinct catalytic activities within its structure which allow it first to generate 5-

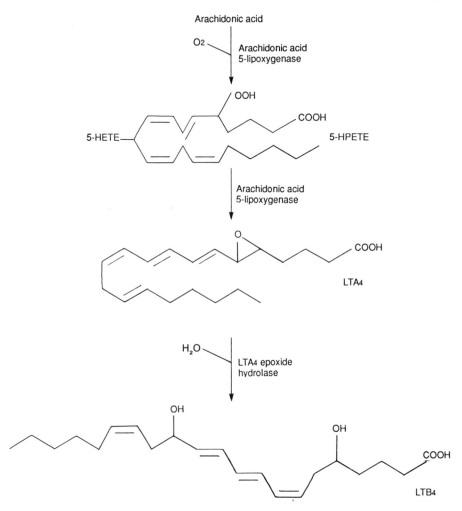

Fig. 7.10 Synthesis of leukotrienes A$_4$ and B$_4$ from arachidonic acid

HPETE from arachidonic acid and then to dehydrate the 5-HPETE in a reaction (Fig. 7.10) which yields LTA$_4$ the leukotriene (LT). These catalytic activities are dependent on the presence of Ca^{2+} ions, are enhanced by ATP and the enzyme has a K_m for arachidonic acid of $\sim 17\,\mu$M.

Investigation of the subcellular distribution of arachidonic acid 5-lipoxygenase has revealed a rather complex pattern which can be altered according to the presence or absence of a number of cellular factors. In general, the enzyme appears to be a soluble protein but its activity can be modulated if membranes are included in the incubation medium. Ammonium sulphate precipitation and ion-exchange fractionation studies have also revealed the existence of two separate activator proteins, either of which can increase 5-

lipoxygenase activity by about 2-fold. The presence of both of these proteins (which remain largely uncharacterized) is required for maximal expression of 5-lipoxygenase and LTA_4 synthase activity. A third protein component which may also contribute to the expression of full enzyme activity is associated with the particulate fraction of human neutrophils, suggesting that the enzyme could be controlled by a translocation mechanism involving adherence to intracellular membranes during the activation process. Indeed, a model has been suggested whereby neutrophil activation leads to a rise in the cytosolic free Ca^{2+} concentration and to a consequent (i.e. Ca^{2+} stimulated) association of the 5-lipoxygenase with a membrane fraction. This translocation then results in the formation of a multimeric protein complex which leads to enzyme activation.

In many cells, LTA_4 can be enzymatically hydrolysed to LTB_4 which is a very potent leucotactic agent and can induce chemotaxis in neutrophils at concentrations as low as 10^{-9} M. Isomers of LTB_4 can also be generated by non-enzymatic hydrolysis of LTA_4, but these do not show biological activity.

LTA_4 can also serve as substrate for the enzyme glutathione-S-transferase which catalyses the covalent attachment of glutathione (Glu–Cys–Gly) to LTA_4 to form a peptide-derivatized molecule, LTC_4. This molecule was shown to be a component of the long-sought SRS-A and is now recognized as the precursor of several other biologically-active molecules which also have bronchoconstrictor activity. These include LTD_4, which is produced by the removal of a γ-glutamyl residue from LTC_4, and LTE_4, which results from peptide cleavage of a glycine residue from LTD_4. Thus, LTC_4, LTD_4 and LTE_4 are all derivatives of LTA_4 that contain a cysteine residue linked through its sulphur atom to the carbon 'backbone' of the fatty acid chain (Fig. 7.11). These molecules can be synthesized by a variety of white blood cells (e.g. neutrophils, monocytes, mast cells) and also by lung, spleen and heart cells.

The enzymes responsible for conversion of LTB_4 to the various cysteinyl derivatives are all membrane-bound proteins and both γ-glutamyl transpeptidase (converting LTC_4 to LTD_4) and dipeptidase (LTD_4 to LTE_4) have been localized to the plasma membrane. The precise subcellular location of the glutathione-S-transferase is less certain, but the fact that it is a particulate enzyme implies that it cannot be the same activity as that resident in most cell types, which is cytosolic. This pattern of distribution, together with the assumption that the targets for leukotrienes C_4, D_4 and E_4 are extracellular (i.e. they are not intracellular second messengers) has been interpreted to suggest that final conversion of LTC_4 to the remaining derivatives may occur as the molecule is about to undergo passage through the plasma membrane. Interestingly, LTC_4 has recently been implicated as a neuroendocrine modulator in the hypothalamic–pituitary system, where concentrations of LTC_4 as low as 10^{-13} M have been reported to stimulate luteinizing hormone secretion from anterior pituitary cells.

The effects of leukotrienes on target cells are mediated by specific receptor molecules which are located at the cell surface and show considerable specificity for individual leukotriene sub-types.

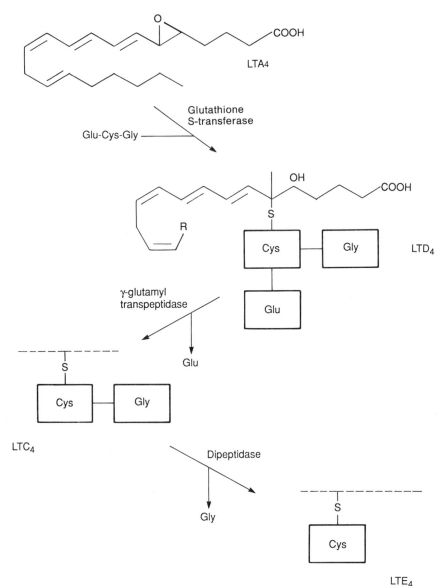

Fig. 7.11   Synthesis of peptide-linked leukotrienes

Lipoxins

Recently, a new lipoxygenase-based pathway of arachidonic acid metabolism has been identified that leads to the formation of two compounds which, like the leukotrienes, are also able to elicit physiological responses when added to neutrophils. These agents do not, however, induce an identical profile of responses when compared with leukotrienes, suggesting that they represent a

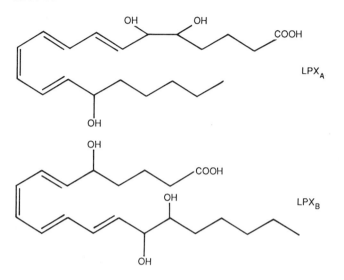

Fig. 7.12 Structures of lipoxins A and B

new class of messenger molecules. Chemical analysis of these compounds revealed that they are tri-hydroxylated derivatives of arachidonic acid that are probably synthesised via intermediate steps involving both the 5- and 15-lipoxygenase enzymes (Fig. 7.12). The biologically active forms have been called lipoxin A and lipoxin B, to denote their synthesis as *Lipox*ygenase *inter*action products.

The biological roles of the lipoxins remain to be established but early evidence suggests that they could have both intra- and extra-cellular targets. Thus, lipoxin A has been found to induce chemotactic responses and to cause smooth-muscle contraction and lipoxin B to inhibit the cytotoxic activity of human natural killer cells. Furthermore, lipoxin A can activate protein kinase-C (see Chapter 8) *in vitro*, and is more potent than diacylglycerol in this respect. While this could clearly represent a physiologically important action, it remains to be determined whether the mechanism operates in intact cells.

It will be evident from the outline provided in this chapter that the single precursor molecule arachidonic acid can be metabolized in a startling number of ways by appropriately equipped cells. Indeed, it seems likely that the complete range of biologically relevant pathways has still not been fully appreciated and that, yet more, active metabolites await discovery. The arachidonic acid cascade, therefore, represents a uniquely versatile signalling system that is used by cells to generate both intra- and extra-cellular second messenger molecules. It is of interest that the derivatives which are active at extracellular sites exert their influence by interaction with receptors coupled to the more 'classical' signalling systems (such as cAMP and Ca^{2+}) and, thus, arachidonic acid metabolites converge with other hormones at this level. Since

PLA$_2$ activity can, itself, be under hormonal control, arachidonic acid metabolism can be viewed as an amplifying and diversifying cascade mechanism that can be directly controlled by, and may functionally interact with, other hormonal mechanisms.

Further reading

Phospholipase A$_2$

Axelrod, J., Burch, R.M. and Jelsema, C.L. (1988). Receptor-mediated activation of phospholipase A$_2$ via GTP-binding proteins: arachidonic acid and its metabolites as second messengers. *Trends Neurosci.*, **1**, 117–23.

Burgoyne, R.D., Cheek. T.R. and O'Sullivan, A.J. (1987). Receptor-activation of phospholipase A$_2$ in cellular signalling. *Trends Biochem. Sci.*, **12**, 332–3.

Chang, J., Musser, J.H. and McGregor, H. (1987). Phospholipase A$_2$: function and pharmacological regulation. *Biochem. Pharmacol.*, **36**, 2429–36.

Fain, J.N. (1988). Evidence for involvement of G-proteins in activation of phospholipases by hormones. *FASEB J.*, **2**, 2569–74.

Irvine, R.F. (1982). How is the level of free arachidonic acid controlled in mammalian cells? *Biochem. J.*, **204**, 3–16.

Jelesma, C.L. and Axelrod, J. (1987). Stimulation of phospholipase A$_2$ activity in bovine rod outer segments by the $\beta\gamma$ subunits of transducin and its inhibition by the α-subunit. *Proc. Natl. Acad. Sci. USA*, **84**, 3623–7.

Kakashima, T. (1988). Stimulation of arachidonic acid release by guanine nucleotides in saponin permeabilized neutrophils: evidence for involvement of a G-protein in phospholipase A$_2$ activation. *Archiv. Biochem. Biophys.*, **261**, 375–82.

Laychock, S.G. (1982). Phospholipase A$_2$ activity in pancreatic islets is calcium-dependent and stimulated by glucose. *Cell Calcium*, **3**, 43–54.

van Kuijk, F.J.G.M., Sevanian, A., Handelman, G.J. and Dratz, E.A. (1987). A new role for phospholipase A$_2$: protection of membranes from lipid peroxidation damage. *Trends Biochem. Sci.*, **12**, 31–4.

Withnall, M.T., Brown, T.J. and Diocee, B.K. (1984). Calcium regulation of phospholipase A$_2$ is independent of calmodulin. *Biochem. Biophys. Res. Commun.*, **121**, 507–13.

Lipocortin

Favvel, J., Salles, J.P., Roques, V. *et al.* (1987). Lipocortin-like anti-phospholipase A$_2$ activity of endonexin. *FEBS Lett.*, **216**, 45–50.

Flower, R.J. (1986). The mediators of steroid action. *Nature*, **320**, 20.

Hayashi, H., Owada, M.K. and Sonaba, S. *et al.* (1987). A 32-kDa protein associated with phospholipase A$_2$-inhibitor activity from human placenta. *FEBS Lett.*, **223**, 267–72.

Hirata, F. (1981). The regulation of lipomodulin, a phospholipase inhibitory protein, in rabbit neutrophils by phosphorylation. *J. Biol. Chem.*, **256**, 7730–3.

Huang, K.S., McGray, P., Mattaliano, R.J. *et al.* (1987). Purification and characterization of proteolytic fragments of lipocortin 1 that inhibit phospholipase A$_2$. *J. Biol. Chem.*, **262**, 7639–45.

Pepinsky, R.B., Sinclair, L.K., Browning, J.L. *et al.* (1986). Purification and partial sequence analysis of a 37-kDa protein that inhibits phospholipase A_2 activity from rat peritoneal exudates. *J. Biol. Chem.*, **261**, 4239–46.

Touquoi, L., Rothhut, B., Shaw, A.M. *et al.* (1986). Platelet activation – a role for a 40-K anti-phospholipase A_2 protein indistinguishable from lipocortin. *Nature*, **321**, 177–80.

Wallner, B.P., Mattaliano, R.J., Hession, C., *et al.* (1986). Cloning and expression of human lipocortin, a phospholipase A_2 inhibitor with potential anti-inflammatory activity. *Nature*, **320**, 77–81.

Arachidonic acid as a second messenger

Beaumier, L., Faucher, N. and Naccache, P.H. (1987). Arachidonic acid-induced release of calcium in permeabilized human neutrophils. *FEBS Lett.*, **221**, 289–92.

Chan, K-M. and Turk, J. (1987). Mechanism of arachidonic acid-induced Ca^{2+} mobilization from rat liver microsomes. *Biochem. Biophys. Acta*, **928**, 186–93.

Kolesnick, R.N. and Gershengorn, M.C. (1985). Arachidonic acid inhibits TRH-induced elevation of cytoplasmic free calcium in GH_3 pituitary cells. *J. Biol. Chem.*, **260**, 707–13.

Kolesnick, R.N., Musacchio, I., Thaw, C. and Gershengorn, M.C. (1984). Arachidonic acid mobilizes calcium and stimulates prolactin secretion from GH_3 cells. *Am. J. Physiol.*, **246**, E458–E462.

McPhail, L.C., Clayton, C.C. and Synderman, R. (1984). A potential second messenger role for unsaturated fatty acids: activation of Ca^{2+}-dependent protein kinase. *Science*, **224**, 622–5.

Morgan, N.G., Rumford, G.M. and Montague, W. (1987). Mechanisms involved in intracellular calcium mobilization in isolated rat islets of Langerhans. *Biochem. J.*, **244**, 669–74.

Wolf, B.A., Turk, J., Sherman, W.R. and McDaniel, M.L. (1986). Intracellular Ca^{2+} mobilization by arachidonic acid. *J. Biol. Chem.*, **261**, 3501–11.

Arachidonic acid metabolism

Bevan, S. and Wood, J.N. (1987). Arachidonic acid metabolites as second messengers. *Nature*, **328**, 20.

Johnson, M., Carey, F. and McMillan, R.M. (1983). Alternative pathways of arachidonate metabolism: prostaglandins, thromboxones and leukotrienes. *Essays Biochem.*, **19**, 40–141.

Lee, J.B. (1982). *Prostaglandins*. New York, Elsevier Science Publishers. 377 pp.

Lewis, R.A. and Austen, K.F. (1984). The biologically active leukotrienes. Biosynthesis, metabolism, receptors, functions and pharmacology. *J. Clin. Invest.*, **73**, 889–97.

Needleman, P., Turk, J., Jakschik, B.A. *et al.* (1986). Arachidonic acid metabolism. *Ann. Rev. Biochem.*, **55**, 69–102.

Piomelli, D., Volterra, A., Dale, N. *et al.* (1987). Lipoxygenase metabolites of arachidonic acid as second messengers for presynaptic inhibition of *Aplysia* sensory cells. *Nature*, **328**, 38–43.

Piper, P.J. (1984). Formation and actions of leukotrienes. *Physiol. Rev.*, **64**, 744–61.

Rokach, J. (1988). The lipoxins. *Int. J. Biochem.*, **20**, 753–8.

Samuelsson, B., Dahlen, S.E. Lindgren, J.P. *et al.* (1987). Leukotrienes and lipoxins. Structures, biosynthesis and biological effects. *Science*, **237**, 1171–6.

Taylor, G.W. and Clarke, SR. (1986). The leukotriene biosynthetic pathway: a target for pharmacological attack. *Trends Pharmacol. Sci.*, **7**, 100–3.

8 Protein phosphorylation in cell regulation

Protein phosphorylation and the control of cell function

It was first established as long ago as 1906 that proteins contain phosphorus, and by 1932 it was already appreciated that cellular proteins can be covalently modified by attachment of phosphate groups to the amino acid serine. It is now clear that by far the majority of protein phosphorylation occurs on either serine or threonine residues (these account for more than 99.9% of all protein phosphorylation sites in cells) while more recent work has revealed that tyrosine residues can also become phosphorylated under certain conditions. The bond that is formed between phosphate and the amino acid involves the oxygen atom which is present in the hydroxyl group of each of the amino acids serine, threonine or tyrosine and, in all cases, the phosphate group is supplied by a nucleoside triphosphate (usually ATP). The resulting link is a stable ester bond which can only be hydrolysed in the cell by the action of a phosphomonoesterase enzyme.

It became apparent during the 1950s that protein phosphorylation occurs in a regulated fashion in cells and that proteins can cycle between one of two forms, the first bearing phosphate groups, the second lacking them. This interconversion process was correlated with changes in the catalytic activity of several enzymes, and led to the concept that reversible alterations in protein phosphorylation state can play a role in the regulation of enzyme activity. The prototype for this kind of metabolic regulation is the control of glycogen metabolism in liver and skeletal muscle. Here, the rate-limiting enzymes for both glycogen breakdown (glycogen phosphorylase) and synthesis (glycogen synthase) are subject to changes in phosphorylation state, which also control their activity. The system is arranged such that they respond in a reciprocal manner to an increase in the extent of phosphorylation. Thus, phosphorylase is activated when it becomes phosphorylated (thereby facilitating glycogen breakdown), whereas glycogen synthase is inactivated by phosphorylation. This mechanism provides economy to the overall regulation of glycogen metabolism since, in principle, activation of a single protein kinase can switch on glycogen breakdown and simultaneously switch off glycogen synthesis. More details of the regulation of this process are given in Chapter 4. It will be evident, however, that such a coordinated system can only function effectively if a number of criteria are satisfied:

1. The separate enzymes that are regulated by any particular protein kinase must all be good substrates for that kinase in order to allow efficient control of both limbs of a pathway. If this is not the case, then it is easy to see that under certain conditions the opposing reactions of a particular pathway could be operating at the same time. For example, if glycogen synthase were not phosphorylated as efficiently as phosphorylase, then glycogen degradation might be initiated under conditions when synthesis was also favoured. This would lead to a potential cycling of glucose from one side of the pathway to the other, with little net output, and would 'damp down' some of the amplification generated by the earlier use of a second messenger.

2. There must be a means to reverse any phosphorylation changes mediated by a kinase, such that hormonal signals can be terminated as rapidly as they are initiated. This is important for the maintenance of sensitivity within cells and allows them to respond rapidly to incoming signals and then, equally rapidly, to re-adjust their function once the hormone concentration outside the cell falls. Any failure at this level would clearly tend to reduce the fidelity of a hormonally controlled system.

The importance of changes in protein phosphorylation state for the regulation of many diverse aspects of cell function is emphasized by the finding that all identified second messenger systems somehow impinge on a protein kinase. In other words, protein kinases are the targets which hormones acting outside the cell are attempting to reach. It is not surprising, therefore, that cells contain an array of these enzymes that each respond to different signals and which often converge on a common substrate as the means to integrate diverse incoming signals. It would not be appropriate here to attempt to provide a complete classification of all protein kinases, but some details relating to the major types of protein kinase are warranted.

Before proceeding to consider the properties of individual protein kinases however, it is also important to briefly examine the effects that phosphorylation can have on a substrate protein. As has already been indicated, in many cases phosphorylation leads to a change in the functional properties of a protein (e.g. enzyme activity) but this is not always the case. When detailed examination of phosphorylated proteins was carried out, it soon emerged that many proteins can be phosphorylated at several different sites, not all of which are functionally equivalent. Indeed, some changes in protein phosphorylation state appear to have absolutely no functional significance at all. For example, the enzyme acetyl-CoA carboxylase is rapidly phosphorylated after treatment of cells with insulin, but this occurs on at least one site in the enzyme which induces no discernible alteration in its activity. Such reactions are referred to as 'silent' phosphorylations.

A well-studied enzyme which is subject to multiple phosphorylation is glycogen synthase. This enzyme has a monomeric M_r of 90 K, but probably exists as a dimer or tetramer under physiological conditions. The enzyme is a good substrate for cAMP-dependent protein kinase (PK-A) which rapidly

Table 8.1 Sites of phosphorylation of glycogen synthase

Kinase	Location of site	Effect on activity
Protein kinase A	1. Near C-terminus	Partial inactivation
	2. Near C-terminus	Complete inactivation
	3. Near N-terminus	None
Phosphorylase kinase	Near N-terminus	Partial inactivation
Ca^{2+}-dependent kinase	Sequential phosphorylation of four sites	1. Increase K_m for UDP-glucose 2–4. Increase in K_a for glucose-6-phosphate

incorporates one phosphate group into the C-terminal region of each subunit. Longer periods of incubation lead to the subsequent phosphorylation of a second C-terminal site, but both sites must be modified in order for complete inactivation of the enzyme to result. However, PK-A can also phosphorylate a third, N-terminal, site on the enzyme with little apparent effect on its activity. Conversely, phosphorylase kinase will also phosphorylate glycogen synthase at a residue near to the N-terminus (although not the same one as at the PK-A site) but this causes partial enzyme inactivation (Table 8.1). Glycogen synthase can also be phosphorylated by a range of other protein kinases, one of which incorporates 4 molecules of phosphate into each subunit of the enzyme, and seems to act in an ordered, sequential fashion. In this case, phosphorylation at the first site causes an increase in the K_m of the enzyme for the substrate UDP-glucose. Modification of the remaining sites is then associated with an increase in the K_a value for glucose-6-phosphate (i.e. more G-6-P is required to activate the enzyme to the same extent). Thus, for this particular kinase, the order of phosphorylation causes different types of change in the substrate enzyme's properties.

Thus, protein phosphorylation can be a very complex process. Multiple sites on a substrate may be targets for a single kinase, and not all of these are necessarily functionally equivalent. Furthermore, different protein kinases can induce similar changes in the properties of a substrate protein by phosphorylating it at separate sites, and, in some cases, the phosphorylation of specific sites may lead to different, but defined, alterations in substrate characteristics.

cAMP-dependent protein kinase

This enzyme is responsible for mediating all of the known effects of cAMP in animal cells and is, therefore, the central regulatory component of all hormonal mechanisms using cAMP as a second messenger. The enzyme is widespread amongst eukaryotic organisms and has been found in fungi and

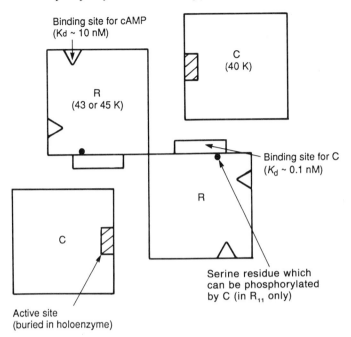

Binding site for cAMP
(Kd ~ 10 nM)

R
(43 or 45 K)

C
(40 K)

Binding site for C
(K_d ~ 0.1 nM)

R

Serine residue which
can be phosphorylated
by C (in R_{11} only)

C

Active site
(buried in holoenzyme)

Fig. 8.1 Diagrammatic representation of the structural features of protein kinase A

higher plants as well as in animal tissues. However, the plant enzyme may be a different molecule from that found in animal cells and, despite the presence of cAMP, there is little evidence that an equivalent activity is present in prokaryotes. All mammalian cells that have been studied contain this enzyme, and it has been found that two distinct forms are present. These isoforms of the enzyme (which are both present in most cell types) can be separated by anion-exchange chromatography and there is some evidence that they may be subject to differential regulation by modified analogues of cAMP. It has also been suggested that they may be subject to selective activation in intact cells by different hormones. However, the mechanism by which this could occur has not been clarified and the possibility must therefore be considered speculative at present. Indeed, since the differences reside in parts of the molecule which are not involved in the catalytic mechanism (i.e. there is no substrate selectivity between the two types) it is not easy to see how such differences could be functionally exploited within cells, unless they play a role in selective compartmentation of the enzymes.

cAMP-dependent protein kinase (PK-A) exists as a tetramer in its holo-enzyme form and is composed of two types of subunit (Fig. 8.1). One of these possesses the catalytic activity of the molecule (and is usually referred to as 'C') and this subunit seems to be identical in both iso-forms of the enzyme. C is composed of about 350 amino acids, has a molecular weight of approximately 40 000 and there are two C-subunits per holoenzyme molecule. To form this molecule, C is complexed with two regulatory 'R' subunits to give an overall R_2C_2 structure. It is the R subunits that are polymorphic and two distinct types have been identified, which are designated R_I and R_{II} respectively. R_I is slightly smaller than R_{II} (43 versus 45 K) and their amino acid sequences have been determined. One major difference between the two isoforms is that R_{II} contains a serine residue which can serve as a substrate for C and can therefore, be phosphorylated, whereas R_I is not a substrate for the enzyme. Most cells contain both types of regulatory subunit, althouth their relative proportions do vary considerably. Indeed, the ratio of the two forms may alter within a given cell type according to particular phases of the cell cycle. Many neonatal tissues contain predominantly type-I PK-A and some evidence suggests that cell transformation can lead to enhanced expression of this enzyme. This, in turn, has fostered the concept that type-I PK-A may have some specialized role in cell growth regulation.

Both forms of the enzyme are located in the cytosolic fraction of cells, although in some tissues (e.g. heart and brain) membrane-associated forms are also present. The membrane-bound enzyme can comprise as much as 50% of the total activity and is predominantly the R_{II} isoform, which seems to bind to cellular membranes *in vitro* more readily than R_I.

Despite differences in primary structure, both types of R subunit share certain common structural features. These include the possession of two high-affinity binding sites for cAMP ($K_d \sim 10\,\text{nM}$) on each R molecule and a second region which participates in a high-affinity interaction with C ($K_d \sim 0.1\,\text{nM}$). The amino acid sequence of the cAMP binding site is very similar in both types of enzyme, whereas those regions of the molecule that are thought to be involved in interaction with C are more diverse between R_I and R_{II}. The serine residue which can be phosphorylated by C occurs at position 95 of the amino acid sequence of R_{II} and lies within the region responsible for binding to C. In addition to interacting with C, each type of R subunit can also interact with a second homologous R subunit to form a dimer. Indeed, this dimeric structure probably represents the normal form of the molecule which occurs in intact cells following release of C. The holoenzyme also contains two C subunits within its structure but these do not interact together in a dimeric fashion, and they are released as independent molecules upon enzyme activation. This occurs when cAMP binds to both of the R subunits. Initially, a ternary complex is formed in which R, C and cAMP all participate but this is unstable due to the development of a dramatic change in the affinity with which R and C interact, resulting from the binding of cAMP. In the presence of cAMP the K_D for interaction between $R + C$ may be increased by several orders of magnitude, resulting in dissociation of the complex, and liberation of the actively catalytic C subunits. Phosphorylation of R_{II} by C may play a role in maintaining this dissociated state since this modification also has the effect of reducing the affinity with which the components interact, and therefore tends to facilitate enzyme activation when cAMP levels are raised. Phosphorylation does not, however, exert as large an influence on the affinity as does cAMP, and it does not directly promote R–C dissociation when cAMP returns to basal levels.

Studies on the kinetics of PK-A activation suggest that phosphorylation of R_{II} may often precede cAMP binding which would therefore, enhance the effectiveness of cAMP to cause subunit dissociation.

Surprisingly, release of C from R does not lead to any increase in the affinity with which substrate ATP is bound by C, but rather it leads to a decrease in this parameter. This indicates that the 'strength' of interaction between C and ATP is not of major importance in determining the enzyme activity. Rather, it means that regulation of the availability of the active site of C is more important in this context. This regulation is achieved through steric interactions, since the presence of R leads to inhibition of the binding of other substrate proteins or peptides by physical blockade of the active site. From this viewpoint, R can be considered to act as a very potent competitive inhibitor of C which binds to, and effectively obscures, the active site. The binding of cAMP to R then induces a conformational change which weakens its interaction with C and results in release of free C. It is likely that R_{II} can serve as a substrate for C even when they are bound together to form the holoenzyme, which emphasizes that, even after phosphorlylation, the substrate (R_{II}) is not automatically released from C without a rise in cAMP concentration.

Substrate specificity

One of the superficially puzzling aspects of the use of a second messenger system such as cAMP is that, while many cells possess the components of the system, each of them responds in a unique way to a rise in cAMP. Therefore, it appears, at first sight, that much of the specificity which arises out of the use of unique receptors control the same enzyme activity (i.e. adenylate cyclase) and therefore converge to use the same biochemical mechanism. However, it is clear that this is not the complete story, as cAMP can induce different responses according to the cell type. Thus, in liver cells it promotes gluconeogenesis and glycogen breakdown, in adrenal cortex it leads to increased steroid output and in many endocrine cells it facilitates an increase in hormone secretion rate. These varied effects must reflect differences in intracellular responses to cAMP and, since all of the actions are mediated by PK-A, it follows that PK-A must act upon different target substrates in each cell type. Thus, the way that a cell responds to a rise in cAMP concentration will be determined exclusively by the available substrates in that cell for PK-A. This, in turn, implies that PK-A cannot be very restricted in its substrate specificity since in different cells it is required to phosphorylate different substrate proteins. Indeed, this prediction can be readily verified if PK-A is allowed to indiscriminately phosphorylate proteins in homogenates of cells. Addition of ATP radiolabelled with ^{32}P in the γ-phosphate position can be used to monitor protein phosphorylation and it soon becomes apparent that in broken cells a very large number of proteins can be phosphorylated by activated PK-A. Caution is required since this does not necessarily imply that all of these are true physiological substrates in the cell, but it emphasizes the potentially broad specificity of the kinase.

Despite the apparently wide range of potential substrates in cells, it is clear that, in general, there are only a small number of true target proteins in any given hormone-sensitive tissue. Some restriction is obviously achieved by appropriate compartmentation of cellular proteins but studies with synthetic substrates have shown that structural criteria are also important. The enzyme will only phosphorylate serine or threonine residues and these have to be appropriately positioned in three dimensions to allow effective interaction with the enzyme's active site. This puts certain constraints on the conformation of potential substrates and it has been established that serine/threonines which reside in regions of proteins that are folded to form α-helix, β-sheet or β-turn structures do not serve as substrates for PK-A. This has led to the idea that a 'random coil' arrangement may be necessary in regions where proteins are to be phosphorylated by PK-A.

In addition to the requirements for appropriate regions of secondary structure (or, perhaps, because of it) a consensus sequence has been identified which represents the primary amino acid sequence favoured by PK-A as a phosphorylation site. This sequence has the structure Arg–Arg–X–Ser–X. Even small peptides bearing this sequence can be phosphorylated by PK-A, which suggests that the more stringent structural constraints which control the folding of larger proteins must play only a rather minor role in determining specificity for the enzyme.

Activation of cAMP-dependent protein kinase A

As mentioned above, it appears that each R subunit of PK-A has two separate binding sites for cAMP and that the overall stoichiometry of the activation process requires the binding of four molecules of cAMP to the R_2C_2 holoenzyme complex. It has been established, on the basis of kinetic criteria, that each of the two cAMP binding sites on each R subunit are not exactly equivalent, and there is a specific order with which cAMP binds to these sites. They have been arbitrarily termed sites A and B, and they are probably not both exposed on the surface of the molecule if no cAMP is bound to R. Dissociation experiments using ^{3}H-cAMP and purified R and C subunits suggest that in the holoenzyme, one cAMP binding site is exposed (site B) and can become occupied if the concentration of cAMP rises. If this occurs, a conformational change results and the second site (A) also becomes exposed on the enzyme surface. Two possibilities then exist: (1) The enzyme can revert to its original state by releasing cAMP from site B, in which case the formation of the intermediate cAMP-bound state is not associated with activation; or (2) cAMP can also bind to the newly exposed site A which then induces subunit dissociation and enzyme activation.

The process of inactivation of the kinase also shows a preferential sequence of events and involves re-association of C with the R subunits containing bound cAMP. Since the whole process is an equilibrium reaction, it is noteworthy that it can be regulated *in vitro* by increasing the concentration of free C subunit. This will facilitate complex formation and will promote R–C association. In the intact cell the frequency of encounters between C and R is also an important determinant of enzyme inactivation rate. It is proposed that the cAMP binding site A lies near the site of interaction between C and R. Hence, binding of C to R is envisaged to promote the dissociation of cAMP from site A and to induce a change in enzyme conformation which also leads to the subsequent release of cAMP from site B. At the same time, this binding reaction also facilitates the loss of cAMP from both sites A and B on the second R subunit of the complex. This overall process, which is illustrated diagrammatically in Fig. 8.2, emphasizes why four cAMP molecules are required for the activation of each molecule of PK-A. Interestingly, it has been found that at concentrations of cAMP similar to those measured in cells under resting conditions, very few molecules of PK-A actually contain no bound cyclic nucleotide. Most are in a state such that one of the subunits contains cAMP bound to site B and the molecule is, therefore, primed to respond to even a minor increase in the prevailing cAMP concentration with an increase in the extent of dissociation and, therefore, of activation. Under resting conditions it has been calculated that approximately 25% of PK-A is present in its dissociated form which suggests that this must represent the basal activity level. Presumably, under these conditions, any changes in protein substrate phosphorylation are compensated by an equally active and opposing phosphatase activity. The basal PK-A activity is also restrained by the presence, in cells, of a specific inhibitor protein which interacts exclusively with free C subunits and prevents substrate phosphorylation. This protein is sufficiently abundant in many cells to act as an effective inhibitor of free C subunits under non-stimulating conditions. It is likely that the rapid increase in protein phosphorylation which accompanies even modest rises in cAMP (about 50% increase in cAMP is often sufficient to cause near maximal activation of a physiological response) reflects the attainment of a threshold value for kinase activation, above which the inhibitor protein is no longer able to cause effective inhibition due to the increased release of free C-subunits.

Another, positive feedback, mechanism may also potentiate PK-A activation in the cell. Cells contain a specific protein (Inhibitor 1) which is a substrate for PK-A and, when phosphorylated, this molecule inhibits the activity of certain protein phosphatases leading to prolongation of the hormonal response. Since PK-A is itself a substrate for its own enzyme activity, and autophosphorylation may facilitate dissociation of R and C, any tendency to prevent dephosphorylation of the kinase will also promote enzyme activation.

These mechanisms combine to produce a highly responsive target enzyme which can respond to very modest changes in cAMP, and can further amplify the response to a hormone by phosphorylating many molecules of substrate before subunit re-association and enzyme inactivation occur.

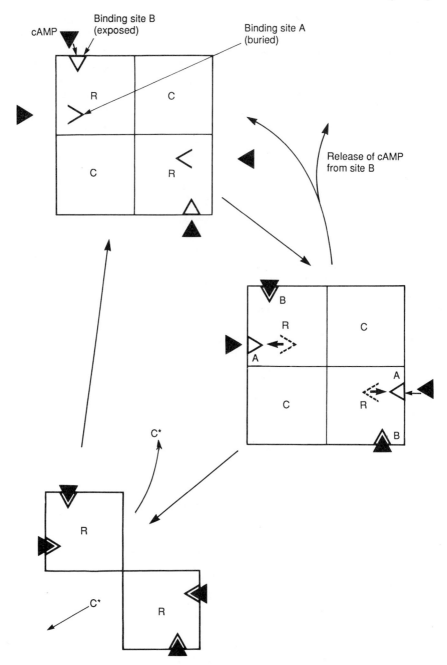

Fig. 8.2 Sequential binding of cAMP by protein kinase A

cGMP-dependent protein kinase

cGMP has been found in many cell types and originally it was seen as a likely second messenger candidate which would be employed by certain hormones in mediation of their cellular responses. Thus, it was envisaged that the cAMP and cGMP systems would be directly comparable and that each might be regulated by different hormones. Unfortunately, this has proved to be an over-simplification and there is little real evidence to support a direct second messenger role for hormone-induced rises in cGMP, except in the case of the recently discovered heart hormone, atrial natriuretic peptide (see Chapter 4). One tissue where a second messenger function for cGMP has been established is the light-sensitive rod cell of the retina. However, here it is a decrease in cGMP, rather than an increase, which results from signal recognition (light absorption) (Chapter 4).

Despite the failure to clarify a generalized role for cGMP in cells, it has been found that many cells contain a protein kinase which can be activated by cGMP. The enzyme is largely cytosolic in distribution and is a dimer composed of two identical subunits (75 000 molecular weight each). Each of these subunits contains a binding site for cGMP and a catalytic domain, and it is proposed that they are arranged asymmetrically such that the catalytic region of one subunit interacts with the regulatory domain of its neighbour, and vice versa (Fig. 8.3). The enzyme is very sensitive to cGMP and can be activated by changes in cGMP concentration which are in the nanomolar range, but activation is not associated with subunit dissociation. Unlike PK-A, PK-G is not susceptible to inhibition by the cellular protein kinase inhibitor protein. cAMP can substitute for cGMP as an activator of the enzyme *in vitro* but a concentration in the micromolar range is required to achieve this effect. Interestingly, the enzyme is subject to autophosphorylation, a response which increases the sensitivity to cAMP by a factor of 10 but has no effect on the

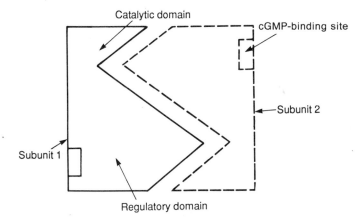

Fig. 8.3 Asymmetric organization of protein kinase G dimers

affinity for cGMP. This could mean that cAMP might possibly regulate the enzyme activity under appropriate conditions inside the cell.

The substrate specificity of PK-G is rather more stringent than that of PK-A, but nevertheless it can also phosphorylate a wide range of peptides and proteins in cell homogenates, and is also subject to autophosphorylation at a particular threonine residue near the N-terminus.

Ca^{2+}-activated protein kinases

Calcium ions have adopted a ubiquitous role as signal agents in cells and, not surprisingly, many of the affects of these ions probably result from activation of protein kinases. It would not be particularly helpful in the present context to simply catalogue these enzymes according to their biochemical properties, but it is appropriate to briefly outline some of their general features. However, one particular Ca^{2+}-sensitive kinase, protein kinase C (PK-C), has begun to assume a special importance in cell signalling, since this enzyme may occupy a position which lies at the convergence of several second messenger systems in the cell. Therefore, more detailed consideration of the properties and function of PK-C will be given below.

Calmodulin-dependent protein kinases

Attempts to isolate Ca^{2+}-dependent protein kinases from different cells rapidly revealed that several types of molecule exist. A number of these derive their sensitivity to Ca^{2+} from an association with the Ca^{2+}-dependent regulator protein, calmodulin (CaM). This small protein acts as the transducer of the Ca^{2+} signal by virtue of a conformational change in the presence of appropriate concentrations of Ca^{2+}, which allows it to interact with, and activate the target kinase. In at least one case (phosphorylase kinase) calmodulin actually forms part of the holoenzyme complex and is present as a subunit of the enzyme in purified preparations. This presumably provides for very efficient kinase activation in response to a rise in cytosolic free Ca^{2+}. Other Ca^{2+}-CaM-dependent enzymes, however, do not have the CaM bound within the enzyme structure. These include enzymes such as myosin light-chain kinase (MLCK) which probably plays a role in agonist-induced smooth-muscle contraction and in the control of motile events in non-muscle cells. This enzyme is itself part of a heterogeneous group, and different tissues may well possess biochemically distinct forms of MLCK. Other CaM-dependent kinases have been purified from liver and brain and include an enzyme often referred to as Ca^{2+}-calmodulin-dependent protein kinase II (PK-II). Ths enzyme has a broad substrate specificity and can phosphorylate similar sites to those which are substrates for PK-A. Skeletal muscle also contains high levels of a more specific and separate enzyme (PK-III) which preferentially phosphorylates a large endogenous protein which remains unidentified in functional terms.

Protein kinase C

Protein kinase C (PK-C) was discovered in 1977 in the laboratory of Dr Y. Nishizuka and was first identified as a protein kinase which can be irreversibly activated by proteolysis. However, it was soon evident that this mechanism of activation is not of physiological significance and that the enzyme is actually a member of the class of Ca^{2+}-activated protein kinases. However, rather than using calmodulin as a co-factor for enzyme activation, PK-C requires the presence of a particular species of acidic phospholipid, usually phosphatidylserine (see below). More strikingly, its activity is also regulated by diacylglycerol (DG). This is important since DG is not normally present to any extent in biological membranes, since it is usually converted to either phosphatidic acid or monoacylglycerol immediately upon formation. However, membrane DG levels can be increased under conditions where phospholipase C activity is raised, suggesting a possible mode of regulation of PK-C. Monoacyl- and triacyl-glycerol cannot induce PK-C activation and free fatty acids are also largely ineffective, although arachidonic acid may be capable of activating the enzyme.

Biochemical properties of protein kinase C

PK-C has been identified in many types of animal cell and is particularly abundant in brain. Originally, the enzyme was assumed to be a single biochemical entity which is of widespread distribution. However, as purification and gene-cloning studies were performed it emerged that at least three distinct types of PK-C exist which show differential tissue distribution. The proteins are all rather similar in size (approximately 670 amino acids) and show considerable sequence homology both at the nucleotide and at the amino acid level, although they can be distinguished immunologically. The three types of PK-C are apparently encoded by entirely separate genes which reside on different chromosomes, and they have been designated α, β and γ. To further confuse the issue, the β-enzyme has recently been shown to exist as two subtypes (β_I and β_{II}) which differ from one another at the C-terminus and which probably arise as a result of the differential processing of a transcript derived from a single gene. This area of research is still very active and it has now emerged that the situation is even more complex than this. For example, three further isozymes of the enzyme have recently been identified from gene cloning experiments, meaning that at least seven different forms of protein kinase C have now been described. Unfortunately, the nomenclature used by different researchers has not yet achieved uniformity but the system suggested by Nishizuka and colleagues, the original discoverers of the enzyme, will be adopted.

Alignment of the amino acid sequences of PK-C α, β and γ reveals that all of the molecules can be divided into nine separate regions. Four of these are highly conserved amongst the different enzymes (regions C1–C4), while the remaining five domains are more variable (V1–V5). The structure is illustrated schematically in Fig. 8.4. One region of the enzyme has been identified that

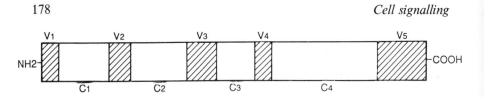

Fig. 8.4 Basic structural organization of the isozymes of protein kinase C

shows marked homology with other protein kinases (including PK-A and PK-G) and is assumed to contain the active site of the enzyme (comprises regions C3–C4). This region is released as a catalytically-active fragment upon proteolysis, which suggests that the regulatory domains of the enzyme are located elsewhere in the sequence, perhaps in regions C1 and C2. If true, then this region of the molecule is likely to be the location of the binding sites for those agents which control the enzyme activity in the cell, e.g. Ca^{2+}, phospholipid and DG.

If the different types of PK-C have specific functions in the cell, then it is likely that they may have somewhat different substrate specificities. This possibility could explain the function of the variable domains of the molecule, which might play a role in determining which substrate proteins can interact with, and be phosphorylated by, PK-C. The precise function of each type of PK-C has not yet been established but it has been found that all four subtypes are simultaneously expressed in adult rat brain, although to varying extents. PK-Cγ appears to be the most restricted in its tissue distribution, being localized exclusively to the central nervous system. It is tempting to postulate that it may, therefore, have a unique role in the control of neuronal activity.

In its native form PK-C is composed of a single polypeptide chain and is an acidic protein (isoelectric point of 5.5). As mentioned above, the enzyme requires phospholipid for activity and the most effective phospholipid for enzyme activation is phosphatidylserine, although phosphatidylinositol will also support activity. Many other membrane lipids, including phosphatidyl-choline, phosphatidylethanolamine and lysophosphatidylserine do not serve as cofactors for activation.

PK-C is reported to be largely present in the cytosolic fraction of unstimulated cells, which does not readily accord with its cofactor and activation requirements, since DG and PS are restricted to cellular membranes. However, stimulation of cells has been correlated with a redistribution of the enzyme, which then becomes associated with the particulate fraction under these conditions. Some caution must, however, be exercised when interpreting such results, since the composition of the medium used for cell disruption can markedly influence apparent distribution patterns. Indeed, recent studies imply that PK-C may actually be associated with membranes, even under resting conditions, but that the extent of this binding is increased following cell activation.

Activation of protein kinase C

In vivo, it is likely that PK-C activity is regulated primarily by the availability

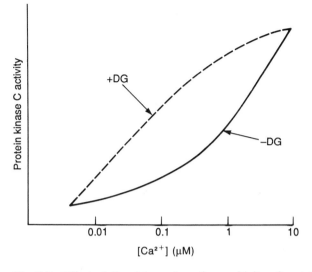

Fig. 8.5 Effect of diacylglycerol on the sensitivity of protein kinase C to Ca^{2+}

of DG rather than by the prevailing Ca^{2+} concentration, although the two may act cooperatively to cause enzyme activation. This makes its mode of activation rather different from that of other Ca^{2+}-dependent protein kinases, whose activity is directly regulated by changes in the concentration of free Ca^{2+}. In the case of PK-C, DG controls the sensitivity of the enzyme to Ca^{2+}, such that an increase in DG generation in the membrane can cause PK-C to become activated, even at $0.1\ \mu M\ Ca^{2+}$, which is close to the resting level in most cells (Fig. 8.5). Hence, while PK-C requires Ca^{2+} for activity, its sensitivity to Ca^{2+} is controlled by DG and it is this factor which probably contributes most to the overall extent of enzyme activation.

The precise structural requirements pertaining to the DG species which most effectively stimulates PK-C have not been extensively investigated. However, it has been established that the identity of the fatty acid at position 1 of the glycerol backbone is not critical, and that DGs containing either long- or short-chain, saturated or unsaturated fatty acids at this position can all activate the enzyme. In contrast, the nature of the substituent at position 2 of the glycerol molecule seems to be more critical. In many cells, the fatty acid esterified at this position is the unsaturated molecule, arachidonic acid (20 carbon atoms and 4 double bonds). The degree of unsaturation may not itself be of crucial importance, but the chain length may be, since as this parameter is decreased the ability of DG to activate PK-C is markedly reduced. The presence of arachidonic acid may be largely coincidental, in that DG with other fatty acids at position 2 can activate PK-C equally well. However, arachidonic acid may serve second-messenger functions of its own in cells (see Chapter 7) and since it can be hydrolysed from DG by the enzyme DG-lipase, this may be of greater value to the cell than its role in allowing DG to activate PK-C.

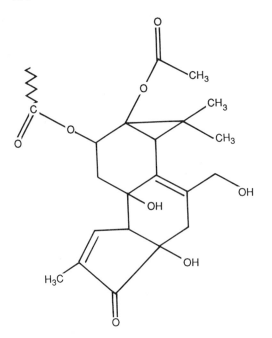

Fig. 8.6   Structure of 12-0-tetradecanoyl-phorbol-13-acetate (TPA)

Phorbol esters

Pharmacologically, PK-C has generated a great deal of recent interest since it has become apparent that this enzyme represents the cellular target for a rather diverse range of compounds which act as tumour promoters. The most potent of these are a group of compounds derived from the diterpene alcohol phorbol which were originally isolated from the oil of the plant *Croton tiglium*, a member of the Euphorbia family. These compounds aroused interest since they are very active as tumour promoters, particularly in skin, and can induce profound alterations in the rate of cell growth and development *in vitro*. The most potent (and most widely studied) tumour-promoting phorbol derivative is a molecule containing two esterified side chains linked to phorbol (a tetradecanoic acid and an acetic acid moiety) and is referred to as 12-O-tetradecanoyl-phorbol-13-acetate (TPA) (Fig. 8.6) or sometimes as phorbol-12-myristate-13-acetate (PMA). Once the structure of TPA was established it became possible to synthesize the molecule (and several derivatives) in radioactive form and to use these to identify possible receptor sites on cells. These studies revealed the presence of a high affinity binding site for TPA in cell membranes ($K_d \sim 500$ pM) which was sensitive to protease treatment and to modulation of the lipid composition of the membrane. In 1982, it was reported by Nishizuka's group that TPA could also activate PK-C and that the manner of this activation resembled that of DG, in that the major effect of the phorbol ester was to increase the Ca^{2+} sensitivity of the enzyme. Furthermore, a range of different phorbol esters showed identical specificity as activators of PK-C and as ligands for the putative cell-membrane receptor. This suggested that these two functional species might be related.

Subsequently, it was possible to demonstrate that both activities (PK-C and phorbol ester receptor) co-purified from mouse brain despite extensive attempts at fractionation using gel filtration, ion-exchange and hydrophobic chromatography. The final product was a single protein (M_r 70 K) which represented pure PK-C and which possessed the appropriate characteristics for the phorbol ester receptor. It is now generally accepted therefore, that PK-C is the cellular receptor for these molecules. Indeed, many studies of the cellular function of PK-C have employed TPA (or structurally-related molecules) as a pharmacological activators of the enzyme.

It should be noted, however, that TPA is a pharmacological tool, and that it may not cause precisely the same changes in cell function that occur when DG levels are increased by a receptor agonist. For example, in many cells, the effects of TPA on PK-C activity are only slowly reversible, whereas, in response to a hormone, the change in enzyme activity is rapidly and completely reversible. Thus, some degree of caution is necessary in the interpretation of data obtained using TPA as an activator of PK-C. Nevertheless, it has been an extremely useful reagent in studies of the function of PK-C in intact cells.

Inhibitors of protein kinase C

An alternative approach for studies of the role of PK-C as a mediator of hormone action is to use inhibitors of the enzyme and to observe their effects on hormone responses in intact cells. In practice, this has not been a particularly fruitful approach since the inhibitors that have been described are not entirely specific. A range of potential candidates is given in Table 8.2, although these should not be viewed as ideal compounds in view of their poor specificity. In this context, it is noteworthy that some of the compounds listed (e.g. phenothiazines, local anaesthetics) were originally thought of as 'specific' inhibitors of calmodulin-dependent processes!

Recently, the possibility that endogenous PK-C inhibitors may exist has been suggested by the partial characterization of two potential inhibitors from brain tissue. Both are proteins, one being quite small with a $M_r \sim 17$ K, while the other is larger, with a $M_r \sim 40$ K. The latter may be dimeric in structure being composed of two monomers, each with a M_r of about 18–20 K. The relationships (if any) between these two molecules and their potential physiological roles as PK-C inhibitors have not yet been determined. It is interesting to speculate, however, that, by analogy with the PK-A system, cells may contain a protein which can inhibit PK-C activity and may, therefore, form part of a feedback control mechanism.

Table 8.2   Inhibitors of protein kinase C

Type of compound	Example
Polypeptides	polymyxin B/melittin
Polyamine	spermine
Antibiotic	doxorubicin (Adriamycin)
Local anaesthetic	tetracaine
Phenothiazine	trifluoperazine/chlorpromazine
Isoquinoline derivatives	'H-7'

Substrate specificity of protein kinase C

Like several other classes of protein kinase, PK-C has a comparatively wide range of potential protein substrates when assayed *in vitro*, not all of which are likely to represent physiological targets *in vivo*. One of the best characterized substrates for the enzyme is a protein that occurs in platelets and which becomes phosphorylated in intact platelets in response to several agonists, including thrombin and collagen. This molecule has a $M_r \sim 40\,K$ and has been implicated in the regulation of serotonin secretion by exocytosis. Other possible substrates for PK-C include the receptor for EGF, the α-subunit of the guanine-nucleotide binding protein G_i and ribosomal protein S6. It is likely that each individual cell type will have a unique set of substrates for the enzyme, and that the availability of particular proteins in the vicinity of the active enzyme will determine precisely which proteins become phosphorylated in any given cell.

Sequence studies with synthetic peptides have revealed that a group of basic residues near to the phosphate acceptor (either serine or threonine) seems to be preferred for activity, although precise details of the constraints have not been established. It is of interest that there is much similarity between the substrate specificity of PK-C and that of PK-A. Indeed, in some cases, these two kinases can both phosphorylate the identical amino acid residue in a protein substrate.

Physiological function of protein kinase C

As indicated above, PK-C becomes activated under conditions when DG generation is increased in the plasma membrane. This will occur when phospholipid breakdown is initiated by hormonal activation of phospholipase C. As described in detail in Chapter 5, this usually results from the hydrolysis of a specialized group of membrane phospholipids: those which contain inositol (or a phosphorylated derivative) as the polar head group. In this case, generation of DG is often accompanied by the production of an isomer of inositol trisphosphate which acts as a second messenger promoting Ca^{2+} release from an intracellular pool. Thus, rises in Ca^{2+} and DG frequently occur at the same time and in response to the same stimulus. This allows for the activation of separate effector mechanisms, among which other Ca^{2+}-activated protein kinases and PK-C will figure prominently. In fact, it has been established that the full expression of a cellular response often depends on the concerted activation of both 'limbs' of this pathway. Hence, while a rise in either Ca^{2+} or DG may, in itself, be a sufficient stimulus to elicit responses such as hormone secretion from endocrine cells, contraction in smooth muscle or neurotransmitter release; when a rise in both Ca^{2+} and DG occur together, the overall response of the cell is much greater than the sum of the individual components, i.e. the two signals synergize.

Since PK-C has been identified as the target for phorbol ester tumour promoters, it follows that the enzyme is also likely to play a role in longer-term growth regulation as well as in acute responses to hormones. These growth-regulating effects may also depend on the Ca^{2+} mobilization pathway, however, since in cells such as lymphocytes, both mechanisms are necessary in order to induce DNA synthesis, even though this takes several hours to be initiated. There is some evidence that growth factors such as insulin, EGF and platelet-derived growth factor also affect cellular mechanisms that are sensitive to PK-C, since these factors often act in concert with tumour promoters to stimulate cell proliferation. However, transcription of certain genes does appear to be activated in response to PK-C alone, although it is not clear whether this is a direct effect of the enzyme at the nuclear level, or whether the intervention of yet another messenger system is required.

One further area of uncertainty which must now be introduced into the PK-C 'story', is the existence of isozymes. At present, nothing is known of the differential properties of the several types of PK-C, but it seems probable that they will be shown to have individual roles to play in the various mechanisms that are regulated by PK-C.

The role of protein kinase C in feedback control

In addition to its role as a positive modulator of cell activation, PK-C also acts to limit the cellular responses to certain stimuli. This is illustrated in the case of the epidermal growth factor (EGF) receptor where PK-C induces phosphorylation of a particular N-terminal threonine residue causing a reduction in the activity of the intrinsic receptor tyrosine kinase activity and lowered binding affinity for EGF. Hence, in this situation, agents that activate PK-C tend to inhibit EGF receptor function. Since EGF does not, itself, cause PK-C activation (stimulation of the EGF receptor does not lead to increased DG production), this mechanism is an example of heterologous feedback regulation, whereby activation of one receptor transduction system results in reduced activity of another, different, system. In hepatocytes, a similar type of desensitization occurs since activation of α_1-adrenergic receptors by catecholamines leads to inositol lipid hydrolysis, DG production and intracellular Ca^{2+} mobilization. The resultant activation of PK-C then leads to attenuation of continued α_1-stimulation since it results in inhibition of further lipid breakdown and thereby reduces the extent of Ca^{2+} mobilization. This mechanism should not be thought of as a feedback system operating at the receptor level (as in the case of agonist-induced receptor down-regulation) since it is not simply restricted to the α_1-system but also leads to attenuation of inositol lipid hydrolysis mediated by both vasopressin and angiotensin II. These peptides act at different receptors to each other and to the α_1-receptor. Thus, in this case, PK-C activation produces down-regulation of a complete signal transducing system rather than simply promoting the uncoupling of a particular set of receptors from that system, as in the case of EGF. It follows from this, that in this case the target protein for active PK-C is unlikely to be the α_1-receptor itself (or any of the other receptors linked to the same transduction system) but must be a component of the signalling system upon which the α_1-receptor impinges. This is emphasized by the finding that prior activation of PK-C also causes inhibition of Ca^{2+}-mobilization in response to $[AlF_4]^-$ in liver, an agent that completely bypasses all of the receptors, but activates a crucial guanine-nucleotide-binding protein which probably form part of the signal mechanism (see Chapter 3). The precise site of action of PK-C has not yet been demonstrated, but the possibility that G-proteins themselves can act as substrates is raised by the observation that G_i (an inhibitory G-protein which regulates adenylate cyclase) can be functionally altered in cells under conditions when PK-C is activated.

Phosphoprotein phosphatases

It is clear from the foregoing part of this chapter that much attention has been focused on the various types of protein kinase that are found in cells and on their potential substrate molecules. Implicit in this preoccupation is the belief that rapid changes in protein phosphorylation state underlie many of the metabolic changes brought about in cells following receptor activation. Accepting the truth of this idea – and the self-evident statement that cells must respond equally rapidly to both increases and decreases in circulating hormone levels – then it is reasonable to suppose that they must also be equipped with active mechanisms to counteract protein kinase activity when stimulation is withdrawn. Since protein kinases catalyze the covalent in-corporation of phosphate into proteins, the primary mechanism for reversing the effects of these enzymes is likely to be one of phosphate removal. Thus, it is not surprising that cells are equipped with several types of phosphatase enzyme that can utilize phosphorylated proteins as a substrate. It is perhaps surprising however, that much less effort has been put into characterizing the phosphoprotein phosphatases than has been expended on the kinases, and our knowledge of their structure and regulation remains relatively scanty. Nevertheless, we should not underestimate the importance of these molecules in hormonal control, nor should we overlook the possibility that they may be directly regulated under certain circumstances. Indeed, it is not uncommon to find that cell activation is associated with both increases and decreases in the extent of phosphorylation of different proteins, suggesting that activation of certain functions may result from protein dephosphorylation. This point deserves emphasis since it is easy to entertain the notion that kinases always activate and phosphatases always inactivate cellular processes. This is patently not true. A prime example can be found in the reciprocal regulation of glycogen metabolism where dephosphorylation inactivates the degradative enzyme phosphorylase, but *activates* the synthetic enzyme, glycogen synthase. Hence, in this pathway, phosphatase action has a direct positive effect to promote glycogen synthesis. Note, therefore, that protein phosphatases (PP) are not (necessarily) simply 'off' switches which counteract a range of positive effects elicited by kinases.

The first PP to be so named was phosphorylase phosphatase (in 1956) although 10 years earlier a report had appeared that protein phosphatase activity was present in amphibian eggs. The coining of the first phosphatase as 'phosphorylase phosphatase' carried the implication of a rather narrow substrate specificity which was later found to be inaccurate, and it remains a fair generalization that these enzymes often have a broad substrate specificity.

With the discovery of increasing numbers of PP much confusion has arisen over the nomenclature of the enzymes. Nevertheless, four major classes have been defined according to the general properties and characteristics of the individual enzymes:

1. ATP, Mg^{2+}-dependent (includes phosphorylase phosphatase)
2. Calcineurin
3. Mg^{2+}-dependent
4. Polycation-stimulated

This classification has been explained by Merlevede and is rather less complicated (though perhaps somewhat less precise) than that used by Cohen which relies on substrate specificity and inhibitor sensitivity and uses differential numbering to distinguish the enzymes.

ATP, Mg^{2+}-dependent protein phosphatase

This enzyme is a broad specificity PP which exists in cells as an inactive holoenzyme complex comprised of two subunits. It is present in most tissues examined and is at particularly high levels in skeletal muscle where it plays a primary role in the regulation of glycogen metabolism. Indeed, the active species (which was first identified as a catalytic subunit of $M_r \sim 35\,K$) was identified as a result of its ability to dephosphorylate, with high activity, the β-subunit of the enzyme phosphorylase kinase. In its native form, the catalytic subunit is associated with a second protein which is usually referred to as 'Inhibitor-2' (I-2). This protein is also of $M_r \sim 35\,K$ and does not normally exist in a free form in cells, since it is all involved in binding to the catalytic subunit. Inhibitor-2 is extremely sensitive to proteolysis which may account for its failure to accumulate in cells when present in an uncomplexed form. A key observation in the understanding of the regulation of this enzyme, was that, in order to generate the active enzyme from the inactive complex with I-2, the latter protein must become phosphorylated at a particular threonine residue. This phosphorylation is catalysed by a kinase, termed phosphatase-1 kinase, whose mode of regulation is not understood. Hence, a prior phosphorylation event is required before active phosphatase activity can be expressed in the cell. This seems an unnecessarily stringent form of regulation, but may represent part of a 'self-regulating' mechanism in which the enzyme dephosphorylates itself autocatalytically as part of the activation mechanism (see below). It is important to note that enzyme activation results from an overall conformation change of the catalytic subunit within the dimeric complex and not from subunit dissociation. Thus, phosphorylation is part of a mechanism which acts to keep the enzyme in an active state. Recent evidence suggests, however, that the phosphorylation step itself may not be the only requirement for production of an active enzyme. Indeed, it has been argued that in its final active form, the complex may already have become dephosphorylated. Thus, the phosphorylation event may be a transient component of the activation process. A model has been proposed whereby the holoenzyme species can exist in one of three different states. Two of these are active and one is an inactive state. Phosphorylation leads to the formation of the first active state by inducing a conformational change such that the catalytic subunit becomes activated. However, it is proposed that under these conditions the active site remains concealed and cannot act upon exogenous substrates, although it is able to dephosphorylate resident I-2. When this occurs (and in the presence of Mg^{2+}) a second conformational change then results which allows other substrates to interact with the newly exposed active site of the molecule. Hence, according to this model (Fig. 8.7), the role of the initial phosphorylation reaction is to promote one change in conformation and to store energy which can then subsequently be used to drive a second conformational change to allow final activation. It is envisaged that the cycle is completed by a spontaneous reconversion to the inactive state.

This model remains somewhat speculative but does provide explanations for several of the unusual features of the activation process.

In the cell, ATP, Mg^{2+}-dependent protein phosphatase (AMDPP) is also subject to regulation by a second protein factor, Inhibitor-1. The mechanism of this inhibition is rather more straightforward than that of I-2, in that I-1 is a good substrate for PK-A and it will only effectively inhibit phosphatase activity in its phosphorylated form. Thus, I-1 serves to amplify a cAMP-dependent signal by restricting phosphatase activity under conditions when PK-A is activated.

In addition to this amplification function, there may also be a second effect of inhibition by I-1, which relates to the fact that AMDPP is a broad specificity enzyme and hence also has substrates that are cAMP-independent phosphoproteins. Since a rise in cAMP will lead to phosphorylation of I-1 and, thereby, to inhibition of AMDPP, this means that cAMP can indirectly modulate the phosphorylation state of proteins that are not regulated by PK-A. Indeed, there is evidence that this may occur in cells. For example,

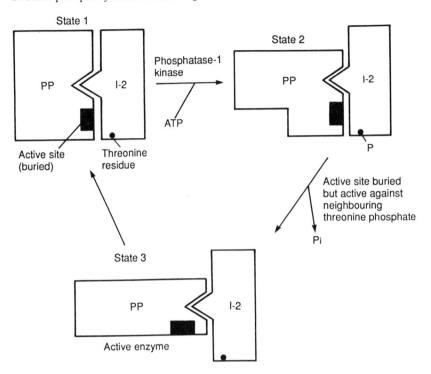

Fig. 8.7 Proposed model for activation of ATP Mg^{2+}-dependent protein phosphatase

adrenaline promotes an increase in the extent of phosphorylation of glycogen synthase in skeletal muscle. However, only part of this effect can be attributed to activation of PK-A, since at least one site on the enzyme whose phosphorylation state is increased under these conditions, is not a substrate for PK-A. It is likely that the increased phosphorylation at this second site reflects inhibition of AMDPP activity.

Recently, yet another potential mechanism by which AMDPP activity could be regulated has also been highlighted. This relates to the tyrosine-specific kinase activity of the viral oncogene product $pp60^{v-src}$. This protein induces transformation of cells infected with Rous sarcoma virus, and is responsible for a significant increase in the extent of tyrosine phosphorylation occurring in such cells. One substrate for this protein kinase (at least *in vitro*) is AMDPP which becomes phosphorylated on tyrosine residues. Significantly, this increase in tyrosine phosphorylation is associated with loss of phosphatase activity, a response that seems to be related to an increase in the K_m of the enzyme for phosphorylated substrates.

It should be noted that, at present, there is no direct evidence to support the idea that AMDPP is phosphorylated on tyrosine residues in intact cells. However, it has been observed that cells transformed with viruses coding for tyrosine kinase enzymes also show enhanced threonine and serine phosphory-lation. One mechanism by which this could occur would be if AMDPP (or another phosphatase) has its activity reduced in such cells, thereby prolonging the effects of a variety of other kinases.

Calcineurin

Calcineurin is a multi-subunit protein phosphatase which is present at high levels in brain and skeletal muscle, although it is also widely distributed in other tissues. Indeed, it is not restricted to mammals as the protein has also been found in the fruit-fly *Drosophila*, suggesting a ubiquitous distribution. The enzyme is dependent on the presence of Ca^{2+} for phosphatase activity and it can interact with calmodulin.

In purified form, the enzyme has been shown to be a dimer composed of two different types of subunit $(A + B)$. The dimer has a total M_r of about 80 K, while the individual subunits are 20 K (B) and 60 K (A). However, different tissues may contain individual isozymes of calcineurin since the A subunit seems to vary in size, according to its source. This may allow the enzyme to dephosphorylate unique substrates in different tissues. Calcineurin has the ability to bind, and to be activated by, Ca^{2+} in the absence of calmodulin, an effect mediated by the B subunit which contains four identifiable Ca^{2+}-binding domains within its amino acid sequence. The binding affinities of these domains are in the micromolar range, suggesting that they could be functional in the intact cell. However, calcineurin also has a high affinity for calmodulin, which binds to the molecule by interaction with regions of the A subunit. Ca^{2+}-calmodulin can increase the activity of the enzyme by up to 10-fold which suggests that calcineurin has a dual mode of regulation by Ca^{2+}, or, put another way, that the enzyme activity is controlled by two separate Ca^{2+}-regulated proteins, one of which forms an integral part of the holoenzyme structure, and the second is calmodulin.

In some tissues (e.g. brain) calcineurin is found in association with the particulate fraction of the cell, suggesting that it may be attached to membranes. This is probably achieved via the B subunit which has been found to contain a myristate substituent near to the N-terminus, and could serve to anchor the protein to the membrane.

The catalytic function of calcineurin seems to reside exclusively in the A subunit since this can be proteolytically cleaved by trypsin to yield a fragment which shows full phosphatase activity but no Ca^{2+} sensitivity. In the cell, it is likely that the enzyme is regulated by the prevailing Ca^{2+} concentration, primarily via a reversible interaction with calmodulin. Calmodulin binds to calcineurin with a stoichiometry of 1:1 and induces enzyme activation by increasing the V_{max} of the reaction without changing the K_m.

Calcineurin shows a rather narrow substrate specificity in that it is most active against sites phosphorylated by either PK-A or PK-G, but it will also dephosphorylate certain proteins that are substrates for Ca-dependent kinases. Since the protein itself is activated by Ca^{2+}, it seems likely that, in the intact cell, it will show preference for sites modified by kinases which are not directly Ca^{2+}-sensitive, and this idea fits well with its reported specificity. The precise regulatory role of calcineurin in any particular tissue has not been established, but its particularly high levels in brain are suggestive of a role in neurotransmitter action.

Mg^{2+}-dependent phosphoprotein phosphatase

This enzyme (MDPP) was first described in liver and heart where it was considered to be involved in the control of glycogen synthase activity. Subsequent studies have revealed that it is widely distributed among various tissues and that it has a very broad substrate specificity. It seems to be particularly effective against the phosphorylated form of the rate-limiting enzyme of cholesterol biosynthesis, hydroxymethylglutaryl-CoA reductase, which may indicate a role in the acute regulation of this pathway in certain cells. One problem with this interpretation is that MDPP is a cytosolic enzyme, while HMG-CoA reductase is microsomal. It is likely that, *in vivo*, MDPP serves a multi-functional role that varies according to the substrate availability in different tissues.

Polycation-stimulated phosphoprotein phosphatase (PSPP)

A protein phosphatase activity that can be activated by very basic proteins was first described in smooth and skeletal muscle, and is now believed to represent a group of enzymes, each with slightly different properties. The activity can be fractionated on DEAE-cellulose ion-exchange columns to yield three distinct species, two of M_r 210 K and another of 150 K. These are all multi-subunit proteins that are believed to possess two common subunits, one of M_r 65 K and a second (the catalytic moiety) of M_r 35 K, while the third subunit is unique to each type of molecule. The available evidence suggests that the proteins are all hetero-trimers and their activity does not appear to be susceptible to modulation by the regulator protein I-1. The three species all show a broad substrate specificity and have the unusual characteristic of being readily activated by cationic proteins, especially histone HI. It does not seem likely that this particular protein can play any physiological role in PSPP activation since it is localized in the nucleus, while the enzymes are probably cytosolic. It is possible that an endogenous cationic activator protein is present in cells but has so far escaped detection. Cellular polycationic molecules that are not proteins might also be potential candidates as activators since synthetic polymers of quaternary ammonium ions can elicit enzyme activation. However, polyamines such as spermine and cadaverine are not effective in this regard. The principal effect of polycations is to lower the K_m of the phosphatase reaction for substrate and this may be the mechanism whereby the enzyme activity is controlled *in vivo*. This may involve subunit dissociation since several *in vitro* studies have revealed that phosphatase activity can be increased after release of the catalytic subunit from the holenzyme. Indeed, one of the forms of the enzyme has been shown to be totally inactive in the trimeric state.

It has been proposed that PSPPs may be involved in the regulation of protein synthesis as they are very effective phosphatases against a phosphorylated form of the eukaryotic initiation factor, eIF-2. Phosphorylation of this molecule is associated with inhibition of the initiation of protein synthesis in cells and phosphatase control of this process would have an obvious potential role in the acute regulation of mRNA translation.

In conclusion, it seems that the phosphatase enzymes present in cells fall into four basic types, which can account for most of the activities presently described. It is likely that these enzymes are as tightly regulated as the kinases whose actions they antagonize, but much more research is still required to provide a clearer understanding of the significance of these enzymes in hormonal regulation, and of the detailed mechanisms by which their activities are controlled.

Regulation of cell function by protein phosphorylation

Activation of steroid synthesis by a cAMP-dependent mechanism

In addition to the regulation of glycogen metabolism in liver and muscle, described in detail in Chapter 4, another tissue in which the second messenger function of cAMP and associated changes in protein phosphorylation can be readily appreciated is the cortical cells of the adrenal gland. These cells represent the site of synthesis of a range of steroid hormones, including the glucocorticoids cortisol and cortisone. Synthesis and secretion of glucocorticoids is controlled by the pituitary hormone, adrenocorticotrophic hormone (ACTH), which binds to a surface receptor on the cells of the adrenal cortex and leads to activation of adenylate cyclase, increased cAMP formation and activation of PK-A. It has been estimated that as much as 1% of the proteins in these cells can serve as substrates for PK-A, although not all of them are likely to be important regulators of steroid production.

A key enzyme in the pathway of steroid synthesis is the enzyme responsible for releasing cholesterol from within the lipid deposits which represent its site of storage in the cells. This enzyme is cholesteryl esterase and it becomes rapidly phosphorylated and activated under the influence of PK-A. The net effect of this is to increase the rate of generation of free cholesterol which is then available for steroid synthesis.

PK-A also acts at a second stage in the process of steroid production, namely at the level of conversion of cholesterol to pregnenolone which is a step common to the synthesis of all steroid hormones. The actual reaction is catalysed by a mitochondrial enzyme complex whose activity is unlikely to be directly controlled by a phosphorylation reaction due to the compartmentali-

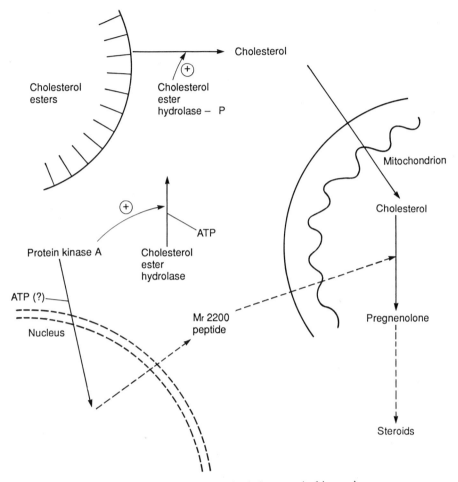

Fig. 8.8 Control of steroid hormone synthesis by protein kinase A

zation of the molecule within an intracellular organelle (where it is inaccessible to PK-A). However, the enzyme activity is regulated by a small cytosolic peptide ($M_r \leqslant 2200$) which promotes pregnenolone synthesis by enhancing the rate of cholesterol interaction with the mitochondrial enzyme complex (Fig. 8.8). The mechanism of this reaction has not been properly defined but it can be readily measured *in vitro* using isolated mitochondria. Interestingly, PK-A activation promotes an increase in the synthesis of this activator suggesting that the adrenal cortex contains some protein substrates which act as regulators of gene expression and, upon phosphorylation, these permit an increase in formation of mRNA for the activator peptide. Hence, this tissue reveals two mechanisms by which protein phosphorylation can be used to elicit cell activation: in one case phosphorylation leads to direct enzyme activation; in the second case it is used to control gene expression. It should be emphasized that the nature and mechanism of action of the proteins involved in the regulation of synthesis of the peptide have not been defined in detail and more work is still required in this area.

Protein phosphorylation and the control of insulin secretion

A system which serves to illustrate the integrated role of protein phosphorylation in controlling cell activation is the regulation of insulin secretion from the B-cells of islets of Langerhans. Even in this case, however, the precise nature of the protein substrates has not been disclosed, although it is possible to make certain reasonable deductions on the basis of considerable circumstantial evidence.

Islet cells contain several types of protein kinase, including PK-A (both isozymes) PK-G, PK-C and Ca^{2+}-regulated kinases such as myosin light chain kinase. The activity of each of these enzymes can be readily observed both in intact cells and in *in vitro* assay systems. Indeed, if islet cells are homogenized and incubated with $[\gamma^{32}P]$-ATP it is possible to demonstrate very extensive phosphorylation of endogenous proteins (not all of which are likely to be of physiological importance!). *In vivo*, the major determinant of the rate of insulin secretion is the prevailing blood glucose concentration, and a rise in glucose is rapidly followed by a corresponding increase in the rate of insulin secretion. At the level of the pancreatic B-cell, glucose leads to activation of all three major intracellular messenger systems since it promotes increases in cAMP (albeit rather modest), cytosolic Ca^{2+} and inositol lipid hydrolysis (with concomitant diacylglycerol production). This leads, in turn, to activation of a number of different protein kinases and to enhanced protein phosphorylation. In practice, it has not proved very easy to identify specific protein kinase substrates in glucose-stimulated islets, which may reflect certain technical problems related to the fact that glucose causes an increase in the rate of ATP turnover in islets and may consequently modify the specific activity of $[^{32}P]$-ATP during the course of an experiment. Notwithstanding these considerations, it has been possible to demonstrate specific protein phosphorylation when the individual kinases are activated more directly. For example, PK-A phosphorylates several islet proteins (M_r 15 K, 55 K and 75 K) as does PK-C (as many as 15 have been identified) and some substrates are apparently common to a number of different protein kinases. Thus, the cells are equipped with an extensive mechanism which can potentially alter the function of a range of endogenous substrates.

In all secretory cells, it is likely that one of the main components of the response mechanism is a change in the rate of movement of the secretory granules from the cytosol to the periphery of the cell. This is necessary since the granules must fuse with the plasma membrane in order to release their contents by exocytosis. It follows, therefore, that a key target for the protein kinases is likely to be either the granule transport system itself, or those proteins which directly control it. In this context it is interesting that islets contain a myosin light chain kinase whose activity is regulated by Ca^{2+} (via

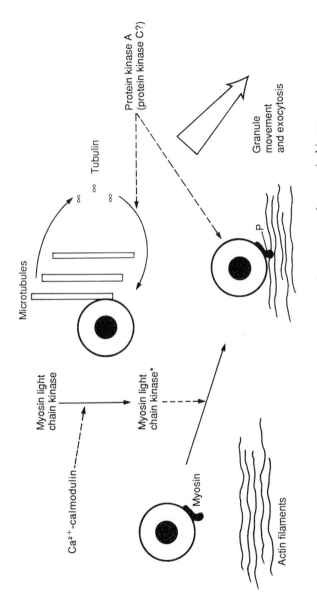

Fig. 8.9 Regulation of secretory granule movement by protein kinases

calmodulin) and that this enzyme controls contractile processes in cells. Furthermore, the cytoskeletal system, comprising both microtubules and microfilaments, plays an important role in secretory granule movement and can be regulated by changes in the protein phosphorylation state. In particular, phosphorylation of the monomeric substituent, tubulin, favours polymerization of microtubules and this phosphorylation can be brought about by both PK-A and by Ca^{2+}-activated kinases. Furthermore, a protein of $M_r = 54\,K$ (which corresponds in size to the β-subunit of tubulin) becomes phosphorylated in glucose-stimulated islets. In addition, the microfilaments are contractile elements composed of actin and their interaction with myosin (which may be present on the secretory granule membrane) is facilitated once myosin has been phosphorylated by myosin light chain kinase.

On the basis of these observations it is possible to construct a working hypothesis which links glucose-induced protein kinase activation with an increase in insulin secretion. In this model (Fig. 8.9), both PK-A and the Ca^{2+}-dependent protein kinases (including PK-C) can be envisaged to share similar target substrates, although the Ca^{2+}-regulated kinases must assume the major role since activation of PK-A alone does not directly stimulate the rate of exocytosis in islet cells, but rather it enhances the secretion rate when another stimulus is also present. Thus:

1. PK-A and Ca^{2+}-dependent protein kinases phosphorylate tubulin and promote microtubule formation.
2. Ca^{2+} also activates myosin light-chain kinase leading to phosphorylation of myosin which is located on the surface of secretory granules.
3. Phosphorylated myosin interacts with microfilamentous actin and forms a contractile system which promotes granule movement.
4. The combined increase in microtubule polymerization and microfilament contraction produces vectorial translocation of secretory granules to the cell surface where exocytosis occurs.

This sequence of events represents a convenient model to explain the regulation of granule movement and it emphasizes how changes in protein phosphorylation can alter both the activity of certain enzymes and can also modulate the functional integrity of the cytoskeletal system by promoting the phosphorylation of structural elements. The details of the process remain to be confirmed experimentally but the process illustrates how different signalling systems can be integrated to produce a coordinated cellular response.

Further reading

Protein kinases – general

Cohen, P. (1988). Protein phosphorylation in hormone action. *Proc. Royal Soc. England.*, **234**, 115–44.

Connelly, P.A., Sisk, R.B., Schulman, H. and Garrison, J.C. (1987). Evidence for the activation of the multifunctional Ca^{2+}/calmodulin-dependent protein kinase in response to hormones that increase intracellular Ca^{2+}. *J. Biol. Chem.*, **262**, 10154–63.

Edelman, A.M., Blumenthal, D.K. and Krebs, E.G. (1987). Protein serine/threonine kinases. *Ann. Rev. Biochem.*, **56**, 567–613.

Granot, J., Mildvan, A.S. and Kaiser, E.T. (1980). Studies of the mechanism of action and regulation of cAMP-dependent protein kinase. *Arch. Biochem. Biophys.*, **205**, 1–17.

Levin, L.R. (1988). A mutation in the catalytic subunit of protein kinase A that disrupts its regulation. *Science*, **240**, 70–4.

Hollenberg, M.D., Valentine-Braun, K.A. and Northup, J.K. (1988). Protein tyrosine kinase substrates: Rosetta Stones or simply structural elements? *Trends Pharmacol. Sci.*, **9**, 63–6.

Hanks, S.K. (1988). The protein kinase family. Conserved features and deduced phylogeny of catalytic domains. *Science*, **241**, 42–52.

Nairn, A.C. and Greengard, P. (1987). Purification and characterisation of Ca^{2+} calmodulin-dependent protein kinase 1 from bovine brain. *J. Biol. Chem.*, **262**, 7273–81.

Nishizuka, Y. (1984). Protein kinases in signal transduction. *Trends Biochem. Sci.*, **9**, 163–6.

Sefton, B.M. and Hunter, T. (1984). Tyrosine protein kinases. *Adv. Cyc. Nuc. Res. Prot. Phos.*, **18**, 195–226.

Scott, J.D., Glaccum, M.B., Zoller, M.J. *et al.* (1987). The molecular cloning of a type II regulatory subunit of the cAMP dependent protein kinase from rat skeletal muscle and mouse brain. *Proc. Natl. Acad. Sci. USA*, **84**, 5192–6.

Taylor, S.S. (1988). Protein kinase A – prototype for a family of enzymes. *FASEB J.*, **2**, 2677–2685.

Voronova, A. and Sefton, B.M. (1986). Expression of a new tyrosine protein kinase is stimulated by retrovirus promoter insertion. *Nature*, **319**, 682–5.

Protein kinase C

Drust, D.S. and Martin, T.F.J. (1985). Protein kinase C translocates from cytosol to membrane upon hormone activation: Effects of thyrotropin-releasing hormone in GH_3 cells. *Biochem. Biophys. Res. Comm.*, **128**, 531–7.

Hidaka, H. and Hagiwara, M. (1987). Pharmacology of the isoquinoline sulfonamide protein kinase C inhibitors. *Trends Pharmacol. Sci.*, **9**, 162–4.

Hirota, K., Hirota, T., Aguilera, G. and Catt, K.J. (1985). Hormone-induced redistribution of calcium activated phospholipid-dependent protein kinase in pituitary gonadotrophs. *J. Biol. Chem.*, **260**, 3243–6.

Hucho, F., Kruger, H., Pribilla, I. and Oberdieck, U. (1987). A 40 kDa inhibitor of protein kinase C purified from bovine brain. *FEBS Lett.*, **211**, 207–10.

Kerr, D.E., Kissinger, L.F., Gentry, L.E. *et al.* (1987). Structural requirements of diacylglycerols for binding and activating phospholipid-dependent Ca sensitive protein kinase. *Biochem. Biophys. Res. Comm.*, **148**, 776–82.

Kikkawa, U. and Nishizuka, Y. (1986). The role of protein kinase C in transmembrane signalling. *Ann. Rev. Cell Biol.*, **2**, 149–78.

Kishimoto, A., Takai, Y., Mori, T. *et al.* (1979). Activation of calcium and phospholipid-dependent protein kinase by diacylglycerol. Its possible relation to phosphatidylinositol turnover. *J. Biol. Chem.*, **255**, 2273–6.

Kraft, A.S. and Anderson, W.B. (1983). Phorbol esters increase the amount of Ca^{2+}, phospholipid-dependent protein kinase associated with plasma membrane. *Nature*, **301**, 621–3.

Kuo, J.F., Schatzman, R.C., Turner, R.S. and Mazzei, G.J. (1984). Phospholipid-sensitive Ca^{2+}-dependent protein kinase: a major protein phosphorylation system. *Mol. Cell Endo.*, **35**, 65–73.

Malaisse, W., Dunlop, M.E., Mathias, P.C.F. *et al.* (1985). Stimulatin of protein kinase C and insulin release by 1-oleoyl-2-acetyl-glycerol. *Eur. J. Biochem.*, **149**, 23–7.

McDonald, J.R. and Walsh, M.P. (1985). Inhibition of the Ca^{2+}- and phospholipid-dependent protein kinase by a novel M_r 17 000 Ca^{2+}-binding protein. *Biochem. Biophys. Res. Comm.*, **129**, 603–10.

Nishizuka, Y. (1984). The role of protein kinase C in cell surface signal transduction and tumour promotion. *Nature*, **308**, 693–8.

Wooten, M.W. and Wrenn, R.W. (1984). Phorbol ester induces intracellular translocation of phospholipid/Ca^{2+} dependent protein kinase and stimulates amylase secretion in isolated pancreatic acini. *FEBS Lett.*, **171**, 183–6.

Protein kinase C – structural aspects

Carpenter, D., Jackson, T. and Hanley, M.R. (1987). Coping with a growing family. *Nature*, **325**, 107–8.

Coussens, L., Parker, P.J., Rhee, L. *et al.* (1986). Multiple, distinct forms of bovine and human protein kinase C suggest diversity in cellular signalling pathways. *Science*, **233**, 859–66.

Jaken, S. and Kiley, S. (1987). Purification and characterization of three types of protein kinase C from rabbit brain cytosol. *Proc. Natl. Acad. Sci. USA*, **84**, 4418–22.

Kikkawa, U., Ogita, K., Ono, Y. *et al.* (1987). The common structure and activities of four subspecies of rat brain protein kinase C family. *FEBS Lett.*, **223**, 212–16.

Kubo, K., Ohno, S. and Suzuki, K. (1987). Primary structures of human protein kinase C βI and βII differ only in their C-terminal sequences. *FEBS Lett.*, **223**, 138–42.

Ohno, S., Kawasaki, H., Imajoh, S., *et al.* (1987). Tissue-specific expression of three distinct types of rabbit protein kinase C. *Nature*, **325**, 161–6.

Ono, Y., Kikkawa, U., Ogita, K. *et al.* (1987). Expression and properties of two types of protein kinase C: Alternative splicing from a single gene. *Science*, **236**, 116–20.

Woodgett, J.R. and Hunter, T. (1987). Isolation and characterization of two distinct forms of protein kinase C. *J. Biol. Chem.*, **262**, 4836–43.

Tumour promoters and activation of protein kinase C

Ashendel, C.L. (1987). The phorbol ester receptor: a phospholipid-regulated protein kinase. *Biochim. Biophys. Acta*, **822**, 219–42.

Blumberg, P.M., Jaken, S., Konig, B. *et al.* (1984). Mechanism of action of the phorbol ester tumour promoters: specific receptors for lipophilic ligands. *Biochem. Pharmacol.*, **33**, 933–9.

Friedman, B., Frackleton, A.R., Ross, A.H. *et al.* (1984). Tumour promoters block tyrosine-specific phosphorylation of the epidermal growth factor receptor. *Proc. Natl. Acad. Sci. USA*, **81**, 3034–8.

Leeb-Lundberg, L.M.F., Cotecchia, S., Lomasney, J.W. *et al.* (1985). Phorbol esters promote α_1-adrenergic receptor phosphorylation and receptor uncoupling from inositol phospholipid metabolism. *Proc. Natl. Acad. Sci. USA*, **82**, 5651–5.

Functions of protein kinase C

Kaibuchi, K., Takai, Y., Sawamura, M. *et al.* (1983). Synergistic funcions of protein phosphorylation and calcium mobilization in platelet activation. *J. Biol. Chem.*, **258**, 6701–4.

Katada, T., Gilman, A.G., Watanabe, Y. *et al.* (1985). Protein kinase C phosphorylates the inhibitory guanine-nucleotide binding regulatory component and apparently suppresses its function in hormonal inhibition of adenylate cyclase. *Eur. J. Biochem.*, **151**, 431–7.

Orellana, S.A., Solski, P.A. and Brown, J.H. (1985). Phorbol ester inhibits phosphoinositide hydrolysis and calcium mobilization in cultured astrocytoma cells. *J. Biol. Chem.*, **260**, 5236–9.

Robinson, J., Badwey, J.A., Karnovsky, M.L. and Karnovsky, M.J. (1984). Superoxide release by neutrophils: synergistic effects of a phorbol ester and a calcium ionophore. *Biochem. Biophys. Res. Comm.*, **122**, 734–9.

Sugden, D., Vanecek, J., Klein, D.C. *et al.* (1985). Activation of protein kinase C potentiates isoprenaline-induced cyclic AMP accumulation in rat pinealocytes. *Nature*, **314**, 359–61.

Volpi, M., Molski, T.F.P., Naccache, P.H. *et al.* (1985). Phorbol 12-myristate 13-acetate pontentiates the action of the calcium ionophore in stimulating arachidonic acid release and production of phosphatidic acid in rabbit neutrophils. *Biochem. Biophys. Res. Comm.*, **128**, 594–600.

Watson, S.P. and Lapetina, E.G. (1985). 1, 2-Diacylglycerol and phorbol ester inhibit agonist-induced formation of inositol phosphates in human platelets: Possible implications for negative feedback regulation of inositol phospholipid hydrolysis. *Proc. Natl. Acad. Sci. USA*, **82**, 2623–6.

Wolf, M., Levine, H., May, W.S. *et al.* (1985). A model for intracellular translocation of protein kinase C involving synergism between Ca^{2+} and phorbol esters. *Nature*, **317**, 546–50.

Yamanishi, J., Takai, Y., Kaibuchi, K. *et al.* (1983). Synergistic functions of phorbol ester and calcium in serotonin release from human platelets. *Biochem. Biophys. Res. Comm.*, **112**, 778–86.

Zawalich, W., Brown, C. and Rasmussen, H. (1983). Insulin secretion: combined effects of phorbol ester and A23187. *Biochem. Biophys. Res. Comm.*, **117**, 448–55.

Phosphoprotein phosphatases – general aspects

Ingebritsen, T.S. and Cohen, P. (1983). Protein phosphatases: properties and role in cellular regulation. *Science*, **221**, 331–7.

Ingebritsen, T.S. and Kohansen, J.W. (1985). Regulation of protein phosphatase 1 by tyrosine protein kinases. *Adv. Prot. Phosphat.*, **1**, 291–308.

Klee, C.C., Krinks, M.H., Manalan, A.S. *et al.* (1985). Control of calcineurin protein phosphatase activity. *Adv. Prot. Phosphat.*, **1**, 135–46.

Li, H.C., Price, D.J. and Tabarini, D. (1985). Subunit composition and regulation of the ATP: Mg(II)-dependent phosphoprotein phosphatase from bovine heart. *Adv. Prot. Phosphat.*, **1**, 39–58.

Merlevede, W. (1985). Protein phosphates and the protein phosphatases. Landmarks in an eventful century. *Adv. Prot. Phosphat.*, **1**, 1–18.

Schlender, K.K., Wilson, S.E., Thysseril, T.J. and Mellgren, R.L. (1985). Characterisation of polycation-stimulated phosphoprotein phosphatases. *Adv. Prot. Phosphat.*, **1**, 311–26.

Phosphorylation reactions in control of cell function

Exton, J.H. (1982). Regulation of carbohydrate metabolism by cyclic nucleotides. In *Handbook of Experimental Pharmacology*, Eds Kebabian, J.W. and Nathanson, J.A., **58**, Springer-Verlag, Berlin.

Harrison, D.E., Ashcroft, S.J.H., Christie, M.R. and Lord, J.M. (1984). Protein phosphorylation in the pancreatic B-cell. *Experientia*, **40**, 1075–84.

Hemmings, H.C., Nairn, A.C., McGuinness, T.L. *et al.* (1989). Role of protein phosphorylation in neuronal signal transduction. *FASEB J.*, **3**, 1583–92.

Howell, S.L. (1984). The mechanism of insulin secretion. *Diabetologia*, **26**, 319–27.

Krebs, E.G. (1985). The phosphorylation of proteins: a major mechanism for biological regulation. *Biochem. Soc. Trans.*, **13**, 813–21.

Simpson, E.R. and Waterman, M.R. (1983). Regulation by ACTH of steroid hormone biosynthesis in adrenal cortex. *Can. J. Biochem. Cell Biol.*, **61**, 692–707.

Glossary of abbreviations

ACTH	adrenocorticotrophic hormone
AMDPP	ATP, Mg^{2+}-dependent protein phosphatase
ANP	atrial natriuretic peptide
ARF	ADP-ribosylation factor
ATP	adenosine triphosphate
C	catalytic subunit of protein kinase A
CaM	calmodulin
cAMP	adenosine $3':5'$ cyclic monophosphate
cGMP	guanosine $3':5'$ cyclic monophosphate
CTP	cytosine triphosphate
DEAE-	dietylaminoethyl-
DG	diacylglycerol
DTT	dithiothreitol
EF-Tu	elongation factor Tu
EGF	epidermal growth factor
eIF-2	eukaryotic initiation factor-2
ER	endoplasmic reticulum
G-protein	guanine nucleotide binding protein
GABA	gamma-amino butyric acid
G6P	glucose-6-phosphate
GppNHp	guanosine imidotriphoshate
GTP	guanosine triphosphate
GTP-γ-S	guanosine $5'$-thio triphosphate
HETE	hydroxyeicosatetraenoic acid
HPETE	hydroperoxyeicosatetraenoic acid
IGF-1	insulin-like growth factor-1
I-1	inhibitor-1
IL-2	interleukin-2
IP	inositol monophosphate
IP_2	inositol bisphosphate
IP_3	inositol trisphosphate
IP_4	inositol tetrakisphosphate
IPG	inositol-phosphate-glycan
LT	leukotriene
MDPP	Mg^{2+}-dependent protein phosphatase
MLCK	myosin light chain kinase
NGF	nerve growth factor

PDGF	platelet-derived growth factor
PDH	pyruvate dehydrogenase
PES	prostaglandin endoperoxide synthetase
PG	prostaglandin
P_i	inorganic phosphate
PI	phophatidylinositol
PIP	phosphatidylinositol-4-phosphate
PIP_2	phosphatidylinositol-4,5-bisphosphate
PK-A	protein kinase-A
PK-C	protein kinase-C
PLA_2	phospholipase A_2
PLC	phospholipase C
PPI	polyphosphoinositides
PS	phosphatidylserine
$R_{(I \& II)}$	regulatory subunit of protein kinase-A
TPA	12-0-tetradecanoylphorbol 13-acetete
TRH	thyrotropin-releasing hormone
TSH	thyroid-stimulating hormone
Tx	thromboxane

Index